Springer Theses

Recognizing Outstanding Ph.D. Research

For further volumes:
http://www.springer.com/series/8790

Aims and Scope

The series "Springer Theses" brings together a selection of the very best Ph.D. theses from around the world and across the physical sciences. Nominated and endorsed by two recognized specialists, each published volume has been selected for its scientific excellence and the high impact of its contents for the pertinent field of research. For greater accessibility to non-specialists, the published versions include an extended introduction, as well as a foreword by the student's supervisor explaining the special relevance of the work for the field. As a whole, the series will provide a valuable resource both for newcomers to the research fields described, and for other scientists seeking detailed background information on special questions. Finally, it provides an accredited documentation of the valuable contributions made by today's younger generation of scientists.

Theses are accepted into the series by invited nomination only and must fulfill all of the following criteria

- They must be written in good English.
- The topic should fall within the confines of Chemistry, Physics and related interdisciplinary fields such as Materials, Nanoscience, Chemical Engineering, Complex Systems and Biophysics.
- The work reported in the thesis must represent a significant scientific advance.
- If the thesis includes previously published material, permission to reproduce this must be gained from the respective copyright holder.
- They must have been examined and passed during the 12 months prior to nomination.
- Each thesis should include a foreword by the supervisor outlining the significance of its content.
- The theses should have a clearly defined structure including an introduction accessible to scientists not expert in that particular field.

Sergio G. Rodrigo

Optical Properties of Nanostructured Metallic Systems

Studied with the Finite-Difference Time-Domain Method

Doctoral Thesis accepted by
The University of Zaragoza, Spain

Springer

Author
Dr. Sergio G. Rodrigo
Departamento de Física de la Materia Condensada
Instituto de Ciencia de Materiales de Aragón
Universidad de Zaragoza
50009 Zaragoza
Spain
e-mail: sergut@unizar.es

Supervisors
Prof. Dr. Luis Martín-Moreno
Departamento de Física de la Materia Condensada
Instituto de Ciencia de Materiales de Aragón
Universidad de Zaragoza
50009 Zaragoza
Spain
e-mail: lmm@unizar.es

Prof. Dr. Francisco José García-Vidal
Departamento de Física Teórica de la Materia Condensada
Universidad Autónoma de Madrid
28049 Madrid
Spain
e-mail: fj.garcia@uam.es

ISSN 2190-5053 e-ISSN 2190-5061
ISBN 978-3-642-23084-4 e-ISBN 978-3-642-23085-1
DOI 10.1007/978-3-642-23085-1
Springer Heidelberg Dordrecht London New York

Library of Congress Control Number: 2011938012

Cover design: eStudio Calamar, Berlin/Figueres

Printed on acid-free paper

Springer is part of Springer Science+Business Media (www.springer.com)

A mis padres

Supervisors' Foreword

The discovery of the laws of electromagnetism (EM) in the nineteenth century triggered an amazing wealth of scientific developments, which have had a profound impact on our society.

Electromagnetism has been developed in many different directions and regimes. However, until recently, the study of electromagnetic fields interacting with objects of size smaller than, but of the order of, the wavelength of the field remained largely unexplored. The reason was the failure, in that regime, of the highly successful approximations that had allowed the development of most of electromagnetic phenomena, namely circuit theory (which applies when scatterers are much smaller than the wavelength) and ray optics (valid when the objects that the field encounters are much larger than its wavelength). Without these tools Maxwell equations were, except in the simplest geometries (presenting a high degree of symmetry, as plane surfaces, spheres...), simply too difficult to handle with existing mathematics.

This represented not only a nagging gap in fundamental science. The present control of sizes and positions of objects in the scale of tens of nanometers has made the understanding of their interaction with light imperative from the technological point of view. Fortunately, computers have evolved very fast and, since the 1990s, are powerful enough both speed- and memorywise to allow solution of Maxwell's equations for many of the basic geometries. Today, this combination of improved manufacturing and computing capabilities is triggering a scientific explosion in what it is now known as the field of Nanophotonics.

Still, the numerical problem is a very difficult one, due to the many different length scales involved, which range from grid sizes of the order of 2–5 nm (needed to describe the penetration of fields in metals) to tens of microns for a small system comprising a few subwavelength objects resonantly coupled.

Nowadays, several computational schemes for solving Maxwell equations have been developed but, due to the inherent complexity of the problem, it is not clear yet which is the best one (or even if there is one that is best for most cases). This thesis focuses on the application of one of the most promising methods, the finite-difference time-domain method (FDTD), to Nanophotonics.

In a nutshell, in FDTD an incident electromagnetic wavefield is propagated in discretized space and discretized time, according to both Maxwell equations and the constitutive relations (which state how materials respond to the EM field). This information is then post-processed to obtain the EM response of the considered system. This method was originally proposed in 1966 by K. Yee, and has been developed over the years, existing now excellent books about (see references in the text). The work presented here closely followed these references. Nevertheless, the actual implementation of a home-made FDTD code still faces some technical problems; the solution to several of them can be found in the text.

The present thesis is, however, not about the FDTD method, but about its application to some physical problems related to the control of EM fields close to metal surfaces. The topics considered include the several aspects on how light transmits through subwavelength apertures in corrugated metal films (such as the influence of the metal, dependence on the metal thickness and the study of optical properties of metal coated microspheres), the optical properties of metamaterials made with stacked hole arrays and the guiding of metallic waveguides (and their focusing capabilities when tapered). These systems are thoroughly analyzed and, whenever possible, the numerical calculations have been accompanied by simplified models that help extract the relevant physical mechanisms at work.

Notably, the thesis also presents many comparisons with experimental data. That this comparison works without the need for a large number of additional fitting parameters is not trivial, as the quality of materials (and thus their optical properties) may, in principle, be altered when these are patterned. The good agreement obtained between experiments and calculations using available data for bulk materials (i.e. without adding fitting parameters) suggests that theory can already be used as a predictive tool in this area.

To summarize, this thesis analyses a large number of topics of current interest in Nanophotonics and the optical properties of nanostructured metals, and presents a short introduction to the FDTD Method. Hopefully, it will be useful both to researchers interested in this numerical method and to those attracted to the field of optical properties of nano- and micro- structured metals.

Zaragoza, Madrid, August 2011

Luis Martín-Moreno
Francisco José García-Vidal

Preface

As everybody has experienced by looking at a mirror, light is almost completely reflected by metals. But they also exhibit an amazing property that is not so widely known: under some circumstances light can "flow" on a metallic surface as if it were "glued" to it. These "surface" waves are called surface plasmon polaritons (SPPs) and they were discovered by Rufus Ritchie in the middle of the past century. Roughly speaking, SPP modes generate typically from the coupling between conduction electrons in metals and electromagnetic fields. Free electrons loose their energy as heat, which is the reason why SPP waves are completely absorbed (in the visible range after a few tens microns). These modes decay through so short lengths that they were considered a drawback, until a few years ago. Nowadays that situation has completely turned. Nano-technology now opens the door for using SPP-based devices for their potential in subwavelength optics, light generation, data storage, microscopy and bio-technology.

There is a lot of research done on those phenomena where SPPs are involved, however there is still a lot of work to do in order to fully understand the properties of these modes, and exploit them. Precisely, throughout this thesis the reader will find a part of the efforts done by our collaborators and ourselves to understand the compelling questions arising when light "plays" with metals at the nanoscale. The outline of the thesis is:

i. Chapter 1: *Introduction*
   First, the fundamentals of SPPs are introduced. In fact, SPPs will be one of the most important ingredients in order to explain the physical phenomena investigated in this thesis.
   Our contributions, from a technical standpoint, have been carried out with the help of two different well known theoretical methods: the finite-difference time-domain (FDTD) and the coupled mode method (CMM). In this chapter, we summarize the most relevant aspects of these two techniques, looking for a better comprehension of the discussions raised along the remaining chapters.

Concerning the rest of experimental and theoretical techniques used, it is out of the scope of this thesis to rigorously describe all of them. Nevertheless, most of those methods, which will not be presented in the introductory chapter, will be briefly explained when mentioned.

ii. Chapter 2: *Extraordinary Optical Transmission*

Imagine someone telling you that a soccer ball can go through an engagement ring. At first, you could think that he or she has got completely mad. A situation like that could have been lived by the researchers who first reported on the extraordinary optical transmission (EOT) phenomenon. Thomas Ebbesen and coworkers found something like a "big" ball passing through a hole several times smaller than it, although there, the role of the ball was played by light. Before Ebbesen's discovery light was not been thought of being substantially transmitted through subwavelength holes. Until 1998, a theory elaborated by Hans Bethe, on the transmission through a single circular hole in a infinitesimally thin perfect conducting screen, had "screened" out any interest in investigating what occurs for holes of subwavelength dimensions. Bethe's theory demonstrated that transmission through a single hole, in the system described above, is proportional to $(r/\lambda)^4$where λ is the wavelength of the incoming light, and r is the radius of the hole. The proportionally constant depends on hole shape, but it is a small number ($\sim$0.24 for circular holes). It is clear that whenever $\lambda \gg r$ transmission is negligible. Nevertheless, Ebbesen and coworkers experimentally found that light might pass through subwavelength holes if they were periodically arranged on a metal surface. More importantly, in some cases even the light directly impinging into the metal surface, and not onto the holes, is transmitted. The SPP modes were pointed to be responsible of EOT.

It is not strange that such a breakthrough sparked a lot of attention in the scientific community. Furthermore, the EOT discovery is not only interesting from the fundamental physics point of view, but from the technological side as well.

The EOT phenomenon strongly depends on both geometrical parameters and material properties. Moreover, EOT does not only occur in two dimensional hole arrays (2DHAs), so other systems have been investigated in the last years. In this way, this thesis is partly devoted to study different aspects of EOT:

(a) We begin by investigating the influence of the chosen metal on EOT using the FDTD method. We analyze transmission spectra through hole arrays drilled in several optically thick metal films (viz. Ag, Au, Cu, Al, Ni, Cr and W) for several periods and hole diameters proportional to the period.
(b) We also study the optical transmission through optically thin films, where the transmission of the electromagnetic field may occur through both the holes and the metal layer, conversely to the "canonical"

configuration where the metal film is optically thick, and the coupling between metal sides can only be through the holes.

(c) On the other hand, since the first experimental and theoretical papers some controversy arose over the mechanisms responsible to enhance optical transmission through an array of holes. Two mechanisms lead to enhanced transmission of light in 2DHAs: excitation of SPPs and localized resonances, which are also present in single holes. In this chapter we analyze theoretically how these two mechanisms evolve when the period of the array is varied.

(d) There are systems displaying EOT different from holey metallic films. One of them is built by monolayers of close-packed silica or polystyrene microspheres on a quartz support and covered with different thin metal films (Ag, Au and Ni). We show that the optical response from this system shows remarkable differences as compared with the "classical" 2DHA configuration.

iii. Chapter 3: *Theory of NRI Response of Double-Fishnet Structures*
Veselago demonstrated that the existence of an isotropic, homogeneous and lineal (i.h.l) medium characterized by negative values of both the permittivity (ε) and the permeability (μ) would not contradict any fundamental law of physics. A substance like that is usually called left-handed material or alternatively, it is said to posses negative refraction index (NRI), and it behaves in a completely different fashion from conventional materials. At the interface between a NRI material and a conventional dielectric medium interesting things would happen. For instance, the current transmitted into a NRI medium would flow through an "unexpected" direction, forced by the Maxwell's equation boundary conditions. Unluckily, no natural material is known to posses a negative value of its refractive index. To date, the only way to achieve NRI materials is by geometrical means. Nevertheless the optical properties of the constituting materials are still important. For instance, as the dielectric constant of metals is "intrinsically" negative, NRI researchers explore how to induce negative permeability on them by designing their geometry in particular ways. This is the reason why these kind of materials are usually called "meta-materials" because their optical response may be different than the optical response of its bulk components. In this chapter we investigate the optical response of one of these metamaterials presenting NRI, a two-dimensional array of holes penetrating completely through a metal-dielectric-metal film stack (double-fishnet structure).

iv. Chapter 4: *Plasmonic Devices*
The special properties of SPPs are being considered for potential uses in circuits. Namely, the possibility of building optical circuits aimed by SPPs has sparked a great interest in the scientific community. As SPPs on a flat surface propagate close to the speed of light, an hypothetical optical SPP-device would be faster than its electronic counterpart. Moreover, different frequencies do not interact, thus several channels would be available for

sending information. A last advantage, SPP-based technology would be compatible to electronic technology since both share the same supporting medium. Transporting optical signals and/or electric ones would be then possible, depending on the characteristics of a specific instrument.

On the contrary, two disadvantages in the use of SPPs instead of electrons arise: **(i)** SPPs are much more difficult to control than electrons on metallic structures (e.g. surfaces), being efficiently scattered by defects present on them, and **(ii)** the finite propagation length of SPP modes. Note that the latter would not be an actual inconvenient in the case of highly miniaturized circuits. Although the SPP modes are well positioned candidates, as we say, they are strongly scattered by any relief on the surface and, due to the mismatch between freely propagating waves and SPPs, they are difficult to be properly excited. A lot of theoretical and experimental works have been devoted on how to guide and generate SPPs.

Regarding the coupling mechanism of light with SPPs, note SPPs can not be excited by an incident plane-wave, because of their evanescent character. There are various coupling schemes that allow light and SPPs to be coupled: prism coupling, grating coupling and near-field coupling. These setups for exciting SPPs are not always useful for certain applications. In Chap. 4 we discuss the advantages and disadvantages of those methods, and we demonstrate a device that enables to create a source for SPPs with remarkable advantages with respect to the other proposals.

In the same chapter we explore different ways for guiding SPP-like modes. Devices for guiding SPPs by means of metallic bumps or holes drilled on a metal surface have been suggested. Another possibility is to guide electromagnetic waves by either a channel cut into a planar surface or a metallic wedge created on it. These structures support plasmonic modes called channel plasmon polarions (CPPs) and wedge plasmon polarions (WPPs) respectively. The surface could be either a metal or a polar dielectric, characterized by negative dielectric constant values. We investigate both CPPs and WPPs by means of rigorous simulations, aimed to elucidate their characteristics, especially, at telecom wavelengths.

We use that information for suggesting a $SPP \leftrightarrow WPP$ conversion device. Lastly we study how gradually tapering a channel carved into a metal surface enables enhanced electromagnetic fields close to the channel apex.

v. Chapter 5: *Optical Field Enhancement on Arrays of Gold Nano-Particles*
Light scattering by arrays of metal nanoparticles gives rise to nanostructured optical fields exhibiting strong and spatially localized field intensity enhancements that play a major role in various surface enhanced phenomena. In general, local field enhancement effects are of high interest for fundamental optics and electrodynamics, and for various applied research areas, such as surface enhanced Raman spectroscopy and microscopy, including optical characterization of individual molecules. Furthermore, the highly concentrated EM fields around metallic nanoparticles are thought to enhance, in turn, non-linear effects, which can pave the way for active plasmonic-

based technologies. Also biotechnology can take advantage of such high intensified optical fields. It is well known that individual metal particles can exhibit optical resonances associated with resonant collective electron oscillations known as localized surface plasmons (LSPs). Excitation of LSPs results in the occurrence of pronounced bands in extinction and reflection spectra and in local field enhancement effects. Such nanoparticles periodically arranged, may cause additional interesting effects. Besides, if nanoparticles are deposited on a metal surface, the emergence of a new channel for light being excited (SPPs) may lead to new phenomena. In this chapter we investigate the optical response of arrays of gold nanoparticles on both dielectric and metal substrates. By means of the FDTD method we analyze the experimental results consisting on: reflection and extinction spectra measuraments along with the non-lineal response known as two-photon excited (photo) luminescence (TPL) generated by inter-band transitions of d-band electrons into the conduction band.

Acknowledgments

I would like to begin by sincerely thanking my supervisors L. Martín-Moreno and F.J. García-Vidal. I didn't only learn theoretical physics from them, but also "experimental" life.

I am also deeply acknowledged people who were involved in those projects that were the seed and the feed of this thesis. It is a long list of collaborators working in the groups of Prof. T.W. Ebbesen, Prof. S.I. Bozhevolnyi, Prof. D. Bäuerle, Prof. J.R. Kreen, Prof. A. Dereux and A.V. Kats at the time the thesis was written. Thanks A. Hohenau, J. Beermann, E. Moreno, A. Mary, L. Landström, F. López-Tejeira, V.S. Volkov and A.Y. Nikitin for your efforts, without this thesis had never been finished.

Acknowledgements

Contents

Abbreviations

AFM	Atomic force microscope
CCOM	Concurrent complementary operators method
CMM	Coupled mode method
CPP	Channel plasmon polariton
DF	Double fishnet
EM	Electromagnetic
EOT	Extraordinary optical transmission
FDTD	Finite-difference time-domain
FFT	Fast fourier transform
FH	Fundamental harmonic
FIB	Focused ion beam
FOM	Figure of merit
FT	Fourier transform
FWHM	Full width at half maximum
LH	Left handed
LIFT	Laser induced forward transfer
LR	Long range surface plasmon polariton
LSP	Localized surface plasmon
MMP	Multiple multipole method
NA	Numerical aperture
NIR	Near infrared
NRI	Negative refractive index
NW	Norton wave
PCS	Photonic crystal slab
PEC	Perfect electric conductor
PLRC	Piece linear recursive convolution method
PML	Perfect matched layer
PS	Polystyrene
PSTM	Photon scanning tunneling microscope
QCM	Quartz crystal microbalance
RH	Right handed

SEM	Scanning electron microscope
SERS	Surface enhanced raman scattering
SIBCs	Surface impedance boundary conditions
SPP	Surface plasmon polariton
SR	Short range surface plasmon polariton
TE	Transverse electric
TM	Transverse magnetic
TPL	Two photon luminescence
UPML	Uniaxial perfect matched layer
VDS	Vacuum-dielectric film substrate
VMDS	Vacuum-metal-dielectric film substrate
WPP	Wedge plasmon polariton
2DHA	Two dimensional hole array

Chapter 1
Introduction

1.1 Electromagnetic Fields Bound to Metals: Surface Plasmon Polaritons

Our investigations have been motivated by the exciting phenomena arising when light interacts with structured metallic systems at the nanoscale. Precisely, most of the physical mechanisms described and investigated in this manuscript result from the interaction of a kind of electromagnetic wave called surface plasmon polariton (SPP) with objects of subwavelength size. In this section, the basic properties of SPP modes are briefly reviewed leaving out the details that can be found elsewhere [1–4], including books on plasmonics [5, 6].

In physics we find plenty of examples that are described by differential wave equations plus a set of boundary conditions. From a mathematical point of view, a confined mode is a solution that exponentially decays far from the defined boundaries. There is a vast number of physical phenomena led by surface modes, but we are interested in those appearing in Plasmonics; the extraordinary transmission of light [7] is a good example.

Much can be understood about an electromagnetic (EM) mode by examining their dispersion relation, i.e., the relationship between the angular frequency (ω) and the in-plane wavevector ($\vec{k}$). This dispersion relationship can be found in different ways; for example, by looking for surface mode solutions of Maxwell's equations under appropriate boundary conditions. We start supposing that an EM wave propagates on the interface between two different media (See Fig. 1.1a) characterized by their respective dielectric constants ($\varepsilon_I, \varepsilon_{II}$). The magnetic permeability μ, is set to be one, which is a good approximation for natural materials at the optical regime. Additionally, it is imposed that this EM wave will propagate along the x-direction, being invariant through the y-direction, thus $\vec{k} = (k_x, 0, k_z^{I,II})$, where $k_z^{I,II} = \sqrt{\varepsilon_{I,II}(\frac{\omega}{c})^2 - k_x^2}$ with $Im(k_z) \geq 0$. Noticeably, as the system is invariant along one of the directions in space, this allows us to distinguish between the two different polarizations. We denote as TM-polarization the one in which the magnetic

S. G. Rodrigo, *Optical Properties of Nanostructured Metallic Systems*,
Springer Theses, DOI: 10.1007/978-3-642-23085-1_1,

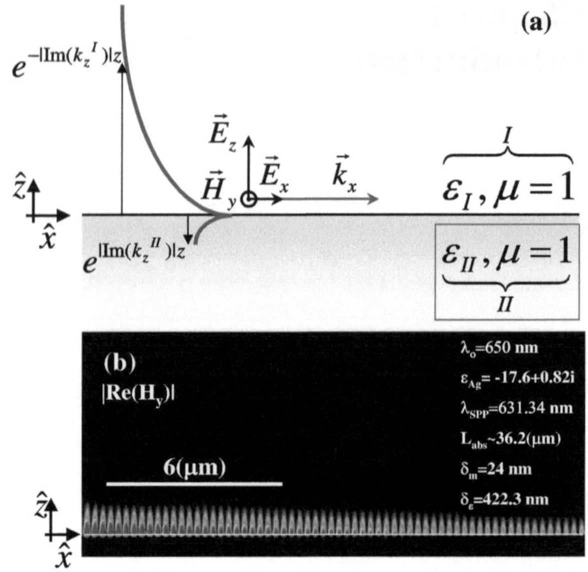

Fig. 1.1 **a** Schematic of the system investigated. **b** Near field representation of $|Re(H_y)|$ for a SPP that propagates on the silver-air interface, being $\lambda_0 = 650$ nm. On the same figure the calculated values of its main defining properties are also shown. (The SPP source (*a magnetic dipole*) is located a few microns from the outer *left*)

field points along the *y*-axis. The other polarization (*TE*) is the one in which the electric field points along the *y*-axis.

For the *TM*-polarization, in region *I*, the magnetic and electric fields are defined as follows,

$$\vec{H}_I = (0, A, 0)e^{ik_x x}e^{ik_z^I z}e^{-i\omega t}$$
$$\vec{E}_I = \frac{-A}{\varepsilon_0\varepsilon_I\omega}(-k_z^I, 0, k_x)e^{ik_x x}e^{ik_z^I z}e^{-i\omega t} \quad (1.1)$$

where A is the amplitude of $\vec{H}_I$. The electric field results from the Maxwell's curl equations (in the MKS system of units):

$$\vec{k} \times \vec{E} = \mu_0\omega\vec{H}$$
$$\vec{k} \times \vec{H} = -\varepsilon\varepsilon_0\omega\vec{E} \quad (1.2)$$

In the same way, the EM fields in region *II* read,

$$\vec{H}_{II} = (0, B, 0)e^{ik_x x}e^{-ik_z^{II} z}e^{-i\omega t}$$
$$\vec{E}_{II} = \frac{-B}{\varepsilon_0\varepsilon_{II}\omega}(k_z^{II}, 0, k_x)e^{ik_x x}e^{-ik_z^{II} z}e^{-i\omega t} \quad (1.3)$$

where B represents the amplitude of $\vec{H}_{II}$. On the surface interface ($z = 0$), boundary conditions impose $(H_x)_I = (H_x)_{II}$ and $(E_x)_I = (E_x)_{II}$, therefore

$$\frac{k_z^I}{\varepsilon_I} = \frac{-k_z^{II}}{\varepsilon_{II}} \quad (1.4)$$

Taking into account the dispersion relation in each medium,

$$
\begin{aligned}
(k_x)^2 + (k_z^I)^2 &= \varepsilon_I \left(\frac{\omega}{c}\right)^2 \\
(k_x)^2 + (k_z^{II})^2 &= \varepsilon_{II} \left(\frac{\omega}{c}\right)^2
\end{aligned}
\tag{1.5}
$$

it can finally be obtained the dispersion relation

$$
k_x = \left(\frac{\omega}{c}\right)\sqrt{\frac{\varepsilon_I \varepsilon_{II}}{\varepsilon_I + \varepsilon_{II}}}
\tag{1.6}
$$

and therefore,

$$
\begin{aligned}
k_z^I &= \pm\left(\frac{\omega}{c}\right)\sqrt{\frac{\varepsilon_I^2}{\varepsilon_I + \varepsilon_{II}}} \\
k_z^{II} &= \pm\left(\frac{\omega}{c}\right)\sqrt{\frac{\varepsilon_{II}^2}{\varepsilon_I + \varepsilon_{II}}}
\end{aligned}
\tag{1.7}
$$

The sign of k_z has to be chosen so that the fields are forced to decay away from the interface, so $Im(k_z^{I,II}) \geq 0$.

By repeating the later process we obtain the condition the *TE* case should fulfill.

$$
k_z^I = -k_z^{II}
\tag{1.8}
$$

As this condition is never satisfied, the *TE*-polarization does not support confined waves. Therefore, as we are searching for EM modes bounded to the surface, the subsequent analysis will go deeply into the *TM*-solution properties.

For the existence of a confined and propagating mode the real part of k_x (Eq. 1.6) must be non-zero, and the imaginary part of both k_z^I and k_z^{II} (Eq. 1.7) must be also different from zero. These conditions ensure that a propagating wave would decay inside both media, as Eq. 1.4 shows. Confinement of EM waves depends on the sign of the real part of the dielectric constant and whether the imaginary part takes different values from zero. Let us consider that medium *I* is a non-absorbing dielectric, in which case $\varepsilon_I = \varepsilon$ is a positive real number. The condition for a surface mode to exist can be obtained from the requirement that the square root expression in Eq. 1.6 has a positive real part, leading to

$$
\begin{aligned}
Re[\varepsilon_I \varepsilon_{II}] &< 0 \\
Re[\varepsilon_I + \varepsilon_{II}] &< 0
\end{aligned}
\tag{1.9}
$$

Note that these conditions are valid whether the imaginary part of ε_{II} is negligible as compared to its real part ($|Re(\varepsilon_{II})| \gg |Im(\varepsilon_{II})|$). According to Eq. 1.9, materials characterized by a negative dielectric constant value may bound an EM mode if it is in contact with a lossless dielectric. Precisely, metals belong to this category.

Before turning to metals, it is interesting to note that also if $Im(\varepsilon_{II}) \neq 0$ EM fields would decay whatever the sign of $Re(\varepsilon_{II})$. When $Re(\varepsilon_{II}) < 0$, such a dielectric constant would describe an absorbing metal. In contrast $Re(\varepsilon_{II}) > 0$ would describe a dielectric material for which absorption has not been neglected. Therefore, the interface between a dielectric without absorption and an absorbing dielectric supports confined modes, usually called Brewster–Zenneck waves [8].

We now return to the case of metals. At optical frequencies (and lower), metals behave like "plasmas", i.e., as if they were gases of free charged particles [9]. The optical response of a free electron gas is approximately described by the Drude model, finding that

$$\varepsilon(\omega) = \varepsilon_r - \frac{\omega_p^2}{\omega(\omega + \imath\gamma)} \tag{1.10}$$

The parameter ε_r gives the optical response at the range of high frequencies, whereas γ is related to energy losses by heating (Joule's effect), and ω_p is the plasma frequency.

Figure 1.2 shows an example. The figure depicts both experimentally measured dielectric constant (circular symbols) and its fit to a Drude-like formula (solid lines). As we can see, the agreement is quite good. Later on (e.g. in Chap. 2) we will see that in order to express accurately the dielectric constant of some metals, additional terms are needed. For the moment, the Drude model contains all the elements required for illustrating the next discussion.

Therefore, if $\varepsilon_I (= \varepsilon)$ is a real positive number and $\varepsilon_{II} = \varepsilon_m$, where the subscript "$m$" states for metals, Eqs. 1.6 and 1.7 define the propagation properties of SPPs.

Figure 1.3 represents the dispersion relation of SPPs on the air-silver interface, where the dielectric constant of silver has been modeled with the Drude parameters appearing in Fig. 1.2. As expected, beyond certain energy values the SPP dispersion relation is clearly distinguished from the light line, a feature due to its intrinsic evanescent character. The anomalous dispersion observed at high frequencies is due to absorption. For lossless metals an asymptotic regime is reached at large wave-vector values. In fact, the SPP frequency tends to $\omega_p/\sqrt{1+\varepsilon_r}$ if the damping coefficient γ is set to zero for the Drude model (Eq. 1.10).

Hereafter we will take a general assumption that is useful for good metals (Ag, Au, Cu), namely that $|\varepsilon'_m| \gg \varepsilon''_m$ ($\varepsilon_m = \varepsilon'_m + \imath\varepsilon''_m$), so $\varepsilon_m \approx \varepsilon'_m$. There are other metals (Al, Ni, Co, Cr, Pb...) for which this approximation is no longer valid, as we will see. In some cases, the condition $|\varepsilon'_m| \gg \varepsilon$ is a good approximation as well.

The properties defining a SPP come from its dispersion relation and the z-component of the $\vec{k}$-vector. These properties tell us what is the spatial "period" of a SPP, how long it takes before being absorbed, and how confined a SPP is inside and outside the metal surface (For a review see [11]). The SPP wavelength is defined as follows,

$$\lambda_{\mathrm{SPP}} = \frac{2\pi}{Re(k_{\mathrm{SPP}})} \tag{1.11}$$

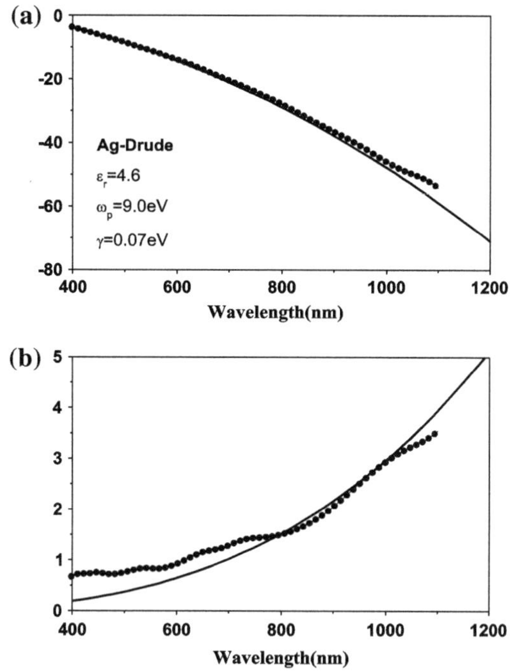

Fig. 1.2 For silver: **a** $Re[\varepsilon_m]$ **b** $Im[\varepsilon_m]$. *Circular symbols* render experimental data [10]. *Solid lines* fit the experiments to a Drude-like formula, defined by the parameters shown in **a**

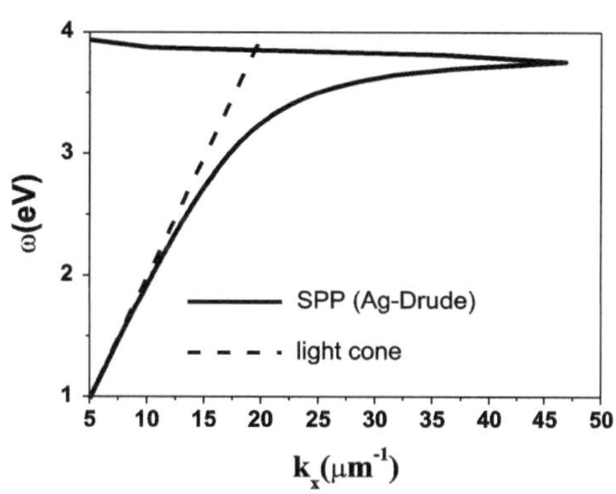

Fig. 1.3 SPP dispersion relation for silver (*solid line*) fitted into a Drude-like formula. We use the parameters shown in Fig. 1.2. The *dashed line* renders the light cone

For good metals, it can be approximated by:

$$\lambda_{\mathrm{SPP}} = \lambda_0 \sqrt{\frac{\varepsilon + \varepsilon'_m}{\varepsilon \varepsilon'_m}} \tag{1.12}$$

where λ_0 is the wavelength in vacuum $\left(\frac{\omega}{c} = \frac{2\pi}{\lambda_0}\right)$. It is easy to see that $\lambda_{\mathrm{SPP}} < \lambda_0$, which it is another consequence of the singular dispersion relation of SPPs (See Fig. 1.3).

The length at which the energy carried by a SPP has decayed a $1/e$ factor is called absorption length and is defined as

$$L_{\text{abs}} = [2Im(k_{\text{SPP}})]^{-1} \tag{1.13}$$

Again, we can make use of the approximation for good metals to obtain

$$L_{\text{abs}} = \lambda_0 \frac{(\varepsilon'_m)^2}{2\pi \varepsilon''_m} \left[\frac{\varepsilon + \varepsilon'_m}{\varepsilon \varepsilon'_m} \right]^{\frac{3}{2}} \tag{1.14}$$

If $|\varepsilon'_m| \gg \varepsilon$, the last formula can be further approximated leading to

$$L_{\text{abs}} = \lambda_0 \frac{(\varepsilon'_m)^2}{2\pi \varepsilon''_m} \tag{1.15}$$

This result means that metals with a large (negative) real part of the relative permittivity are better for guiding or for resonant processes (which require long time to occur). It clearly shows the role played by the damping factor of metals in the SPP behavior: $L_{\text{abs}} \to \infty$ when the imaginary part of the dielectric constant (ε''_m) tends to zero, i.e., as the damping goes to zero too.

Interestingly, for good metals the SPP electric field is primarily transverse in the dielectric and longitudinal in the metal, as the following expressions demonstrate,

$$|E_z^{\varepsilon}| = \sqrt{\frac{|\varepsilon'_m|}{\varepsilon}} |E_x|, \quad |E_z^m| = \sqrt{\frac{\varepsilon}{|\varepsilon'_m|}} |E_x| \tag{1.16}$$

showing the hybrid nature of SPPs that combines the features of both propagating EM waves in dielectrics and free electron oscillations in metals. Since the SPP damping occurs due to ohmic losses ($\sim \vec{j}\vec{E}$), which in metals is related to the charge current ($\vec{j}$) induced by the SPP fields, it is the longitudinal electric field component (E_x) of the SPP in the metal that determines absorption.

It is worth defining another magnitude which can deliver useful information about the SPP nature: the penetration of the SPP fields into each medium. In the dielectric half-space it takes the form $\delta_{\varepsilon} = [Im(k_z^{\varepsilon})]^{-1}$ and in the metal, where it is called skin depth $\delta_m = [Im(k_z^m)]^{-1}$. For lossless metals, skin-depth formulas can be rewritten in a compact manner,

$$\begin{aligned} \delta_m &\approx \frac{\lambda}{2\pi \sqrt{|\varepsilon'_m|}} \\ \delta_{\varepsilon} &\approx \frac{\sqrt{|\varepsilon'_m|}\lambda}{2\pi \varepsilon} \end{aligned} \tag{1.17}$$

The penetration depth of the field into the dielectric gives us a measure of the length scale over which the SPP mode is sensitive to the presence of changes in refractive index, for example the presence of certain bio-molecules in a biosensor.

If we substitute in Eq. 1.17 the expression of ε_m using the Drude formula ($\gamma \sim 0$), and noting that we are working well below the plasma frequency ($\omega \ll \omega_p$) one obtains for the penetration length into the dielectric

$$\begin{aligned} \delta_\varepsilon &= \frac{\lambda^2}{2\pi\varepsilon\lambda_p} \\ \delta_m &= \frac{\lambda_p}{2\pi} \end{aligned} \tag{1.18}$$

where $\omega_p = 2\pi/\lambda_p$. Values for ω_p are around ~9 eV, i.e., $\lambda_p \sim 137.7$ nm, so in this case, the confinement of a SPP could be considered subwavelength up to ~865 nm, since $\delta_\varepsilon < \lambda$ for shorter wavelengths. On the other hand, it is interesting that the penetration depth in metals depends rather weakly on the wavelength, staying at the level of a few tens of nanometers ($\delta_m \sim 22$ nm), while that in dielectrics increases fast and nonlinearly with the wavelength. The penetration depth into the metal gives us a measure on the required metal thickness that allows coupling to freely propagating light in the prism coupling (Kretschmann) geometry (typically 50 nm for silver and gold in the visible). It also sets the length scale of the film thickness so that direct transmission through the film occurs. Moreover, the skin depth gives information about the coupling strength between SPPs at opposite sides of the film. The penetration depth into metals also gives us an idea of the feature sizes needed to control SPPs: as features become much smaller than the penetration depth into the metal they will have a diminishing effect on SPP modes. In SPP investigations, the small-scale (nm) roughness is associated with many of the fabrication techniques that create the metal films. Due to this, a minor perturbation to the SPP mode is provided.

All these quantities (λ_{SPP}, L_{abs}, δ_m, δ_ε) have been represented in Fig. 1.4 for two different metals: silver [panels (a) and (b)] and nickel [Panels (c) and (d)]. Nickel is considered a "bad" metal due to the huge imaginary part of its dielectric constant. We can observe for both metals that at long wavelengths $\lambda_{\mathrm{SPP}} \to \lambda_0$, as Eq. 1.12 predicts. As we said, the imaginary part of ε_m is greater for Ni than for Ag, which explains the differences between the calculated values of L_{abs}. Nevertheless their skin depths are similar. As the figure clearly shows, the approximations that have led to approximated values for δ_m and δ_ε are no longer valid in the case of "bad" metals, as one could expect.

1.2 The Finite-Difference Time-Domain Method

1.2.1 The FDTD Algorithm

The finite-difference time-domain (FDTD) method belongs to the general class of grid-based differential time-domain numerical methods. The time-dependent

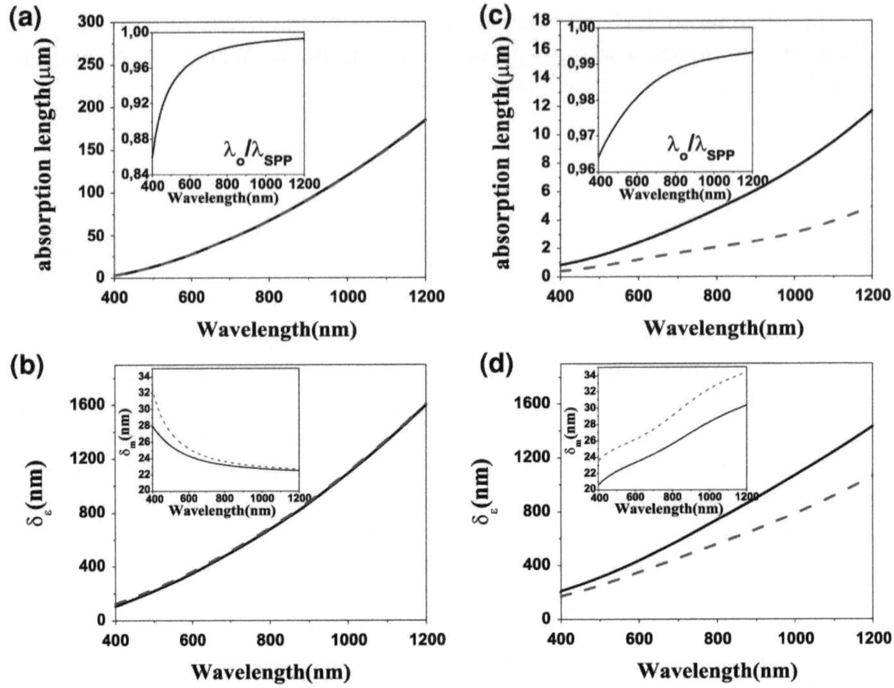

Fig. 1.4 Characteristics of a SPP on the air-silver interface, **a** shows the SPP absorption length for silver (described by a *Drude term*). Inset: ratio between the wavelength of light and the SPP one. Additionally, the main figure in **b** depicts with *solid line* the SPP skin-depth in air. Inset: SPP skin-depth into the metal. *Dashed lines* render their approximated values (Eqs. 1.15 and 1.17). **c** and **d** are as **a** and **b** but for SPPs on the air-nickel interface

Maxwell's equations (in partial differential form) are discretized using central-difference approximations to the space and time partial derivatives. Both the basic FDTD space grid and the time-stepping algorithm trace back to a seminal 1966 paper by Kane Yee [12]. The resulting finite-difference equations are solved in a leapfrog manner: the electric field vector components in a volume of space are solved at a given instant in time; then the magnetic field vector components in the same spatial volume are solved at the next instant in time; and the process is repeated over and over again until the desired transient or steady-state electromagnetic field behavior is fully evolved.

Note that the FDTD technique is one of the most extensively developed and used in computational electromagnetism [13]. It is now impossible trying to cover all aspects of the FDTD method in an introductory chapter. Hence this section is not intended to be a complete FDTD guide, instead, our intention is to give the reader a summarized version of the FDTD method. We will emphasize those techniques that were developed in the course of the thesis and which, to our knowledge, can not be found in the literature. Although these technical issues have not been fully explained in our articles, they were of the utmost importance for achieving the objectives therein.

To start with, we recall some of the most important benefits on the use of the FDTD method:

i. Different sort of material properties can be treated with FDTD, so we are able to properly deal with dielectrics, metals, non-linear substances...
ii. There are a lot of available illuminating sources, for instance: plane waves, dipole sources, gaussian beams...
iii. It is easy to retrieve the optical properties that describes the physical response of a system: transmission and reflection coefficients, points at dispersion relation curves, field maps in the frequency domain or whatever quantity depending upon the EM fields.
iv. This method is fast and it does not consume excessive computer resources compared with other numerical methods.

Let us turn to the FDTD algorithm itself. The starting point are the curl Maxwell's differential equations for isotropic, homogeneous and lineal (i.h.l.) media (MKS system of units)

$$\begin{aligned}\frac{\partial \vec{H}(\vec{r},t)}{\partial t} &= -\frac{1}{\mu_0\mu}\nabla \times \vec{E}(\vec{r},t)\\ \frac{\partial \vec{E}(\vec{r},t)}{\partial t} &= \frac{1}{\varepsilon_0\varepsilon}\nabla \times \vec{H}(\vec{r},t)\end{aligned} \tag{1.19}$$

Note that in principle there is not only a way to bring Maxwell's equations from the "continuous" to the "discrete" space. In the end, the really important question is whether the scheme used for, gives accurate results being free of divergences, numerical instabilities,... The FDTD method is one among other possibilities to solve numerically the curl Maxwell's equations. When Maxwell's differential equations are examined, it can be seen that the change in the E-field in time (the time derivative) is dependent on the change in the H-field across space (the curl), and viceversa. Figure 1.5 shows an illustration of a standard Cartesian Yee's cell used for FDTD, and how electric and magnetic field vector components are distributed [12]. Visualized as a cubic box, the electric field components form the edges of the cube, and the magnetic field components form the normals to the faces of the cube. A three-dimensional space lattice is comprised of a multiplicity of such Yee cells. A given structure is mapped into the space lattice by assigning appropriate values of permittivity to each electric field component, and permeability to each magnetic field component. Yee's scheme proposes a distribution in space for the EM field components. We will see that this leads to an algorithm for the spatial dependence. However each Maxwell's curl equation is coupled to each other, so it is not straightforward to decide the time-stepping. At any point in space, the updated value of the H-field in time is dependent on the stored value of the H-field and the numerical curl of the local distribution of the E-field in space. Yee found that the iteration of E-field and H-field updates results

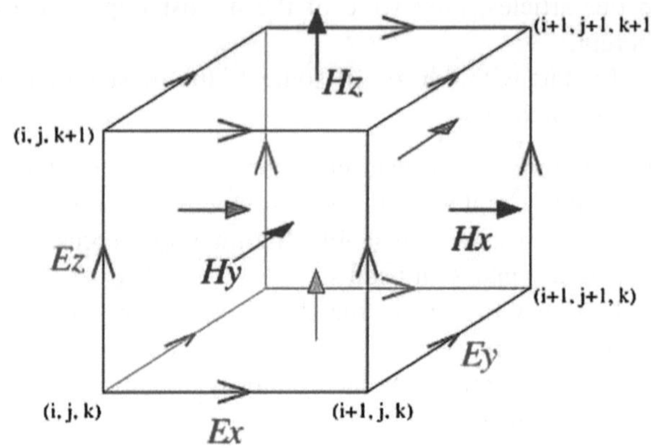

Fig. 1.5 Illustration of a standard Cartesian Yee cell used for FDTD, about which electric and magnetic field vector components are distributed

in a marching-in-time process, i.e., the electric field at time t depends on the electric field at $t-\delta t$ and the magnetic field (via the curl) at $t-\delta t/2$ (δt is the time step). Once the electric field at time t is know the process is iterated, this time in order to solve the magnetic field at time $t+\delta t/2$, which in turn depends on $H(t-\delta t/2)$ and $E(t)$. The last is usually called "leapfrog" algorithm.

Let us briefly show how the basic FDTD algorithm is obtained. The integral form of the Faraday's and Ampere's laws are the best way to get it,

$$\begin{aligned}\frac{\partial}{\partial t}\int \vec{H}(\vec{r},t)d\vec{s} &= -\frac{1}{\mu_0\mu}\oint \vec{E}(\vec{r},t)d\vec{l}\\ \frac{\partial}{\partial t}\int \vec{E}(\vec{r},t)d\vec{s} &= \frac{1}{\varepsilon_0\varepsilon}\oint \vec{H}(\vec{r},t)d\vec{l}\end{aligned} \tag{1.20}$$

As we see in Fig. 1.5 each component of the $\vec{E}$ field can be viewed as surrounded by a circulating current of $\vec{H}$ components, and viceversa. Precisely the EM field component perpendicular to a given face of the Yee's cell represents its averaged value on that surface. Interestingly, there is a connection between Yee's discrete space and the simplest discretization of Faraday's and Amperes's laws in its integral form.

Let us apply Faraday's law to one of the Yee's cell faces in order to calculate H_y. The left hand side reads

$$\frac{\partial}{\partial t}\int \vec{H}(\vec{r},t)d\vec{s} \approx \Delta x\Delta z\frac{\partial}{\partial t}\left[H_y|_{i+\frac{1}{2},j,k+\frac{1}{2}}\right] \tag{1.21}$$

and the right side (counterclockwise integration),

$$\begin{aligned}-\frac{1}{\mu_0\mu}\oint \vec{E}(\vec{r},t)d\vec{l} \approx \frac{1}{\mu_0\mu}&\left\{\Delta x\left[E_x|_{i+\frac{1}{2},j,k+1} - E_x|_{i+\frac{1}{2},j,k}\right]\right.\\ &\left.+\ \Delta z\left[E_z|_{i,j,k+\frac{1}{2}} - E_z|_{i+1,j,k+\frac{1}{2}}\right]\right\}\end{aligned} \tag{1.22}$$

thus,

$$\frac{\partial}{\partial t}\left[H_y|_{i+\frac{1}{2},j,k+\frac{1}{2}}\right] = \frac{1}{\mu_0\mu}\left\{\frac{E_x|_{i+\frac{1}{2},j,k+1} - E_x|_{i+\frac{1}{2},j,k}}{\Delta z} + \frac{E_z|_{i,j,k+\frac{1}{2}} - E_z|_{i+1,j,k+\frac{1}{2}}}{\Delta x}\right\} \tag{1.23}$$

The "leapfrog" algorithm alternates the update of E-fields and H-fields as explained. This translates into the FDTD notation as $\frac{\partial \vec{E}}{\partial t} \approx \frac{\vec{E}^{n+1}-\vec{E}^{n}}{\Delta t}$ and $\frac{\partial \vec{H}}{\partial t} \approx \frac{\vec{H}^{n+1/2}-\vec{H}^{n-1/2}}{\Delta t}$. So finally the H_y update is:

$$H_y|^{n+\frac{1}{2}}_{i+\frac{1}{2},j,k+\frac{1}{2}} = H_y|^{n-\frac{1}{2}}_{i+\frac{1}{2},j,k+\frac{1}{2}} + \frac{\Delta t}{\mu_0\mu}\left\{\frac{E_x|^n_{i+\frac{1}{2},j,k+1} - E_x|^n_{i+\frac{1}{2},j,k}}{\Delta z} + \frac{E_z|^n_{i,j,k+\frac{1}{2}} - E_z|^n_{i+1,j,k+\frac{1}{2}}}{\Delta x}\right\} \tag{1.24}$$

The rest of the electric and magnetic vector components, can be found straightforwardly following this scheme. Once all the components are calculated, this "piece" of algorithm allows us to simulate propagation of EM waves through i.h.l media, defined by the dielectric constant ε, and the magnetic permeability μ.

For the topics covered in this thesis this "particular" FDTD algorithm is of limited interest by itself, since it does not work with metals. However, it appears everywhere in our codes because the studied systems are always embedded in "vacuum" regions. The algorithm must satisfy some criteria in order to be numerically stable. For instance, it can be demonstrated that the dispersion relation for a freely propagating plane wave through the discretized "vacuum" space holds the following formula [13]:

$$\frac{4}{\Delta t}\sin^2\left(\frac{\omega\Delta t}{2}\right) = 4c^2\left[\frac{1}{\Delta x}\sin^2\left(\frac{k_x\Delta x}{2}\right) + \frac{1}{\Delta y}\sin^2\left(\frac{k_y\Delta y}{2}\right) + \frac{1}{\Delta z}\sin^2\left(\frac{k_z\Delta z}{2}\right)\right] \tag{1.25}$$

This result implies that the non-dispersive "continuous" free space becomes dispersive when Maxwell's equations are defined into the time-space lattice. Clearly, in the limit of both mesh size and time step going to zero, the dispersion relation in free space is recovered.

From the dispersion relation, by preventing ω from being a complex number, i.e., $|\sin^2\left(\frac{\omega\Delta t}{2}\right)| \leq 1$ the stability criterium is obtained:

$$(\Delta t)^2 < \left(\frac{c^2}{\Delta x^2} + \frac{c^2}{\Delta y^2} + \frac{c^2}{\Delta z^2}\right)^{-1} \tag{1.26}$$

The last expression implies that we can not independently choose the mesh size and time step. Once the mesh size has been fixed, the time step must be such that the criterium of stability is fulfilled. For a given structure, the mesh size will additionally depend on two important constrains:

i. When the structure to be simulated can not be exactly accommodated in cartesian coordinates, the mesh size should be fine enough to ensure that the discrete structure represents the actual one.
ii. We must take into account the way EM fields are described in the FDTD algorithm. In the case of metals, the EM field decays in length scales of the order of 25 nm. The faithful representation of such fast variations is a great challenge, forcing the mesh size to be usually smaller that 5 nm.

1.2.2 Field Sources in FDTD

Up to here, we have been devoted to show the basics of the FDTD algorithm. However, the algorithm by itself it is not enough. If the EM fields on the grid at time $t = 0$ have not been defined, we will get a lot of zeroes as output after iterating the FDTD loop. The subject of sources for FDTD is one of the most challenging in this theoretical framework. Sometimes it is very difficult to find the proper way to illuminate a structure. For instance, in two-dimensional (2D) periodic systems at normal incidence, it is very easy to use a wave-packet (e.g. gaussian beam), while a monochromatic wave requires further efforts. Illumination by a plane-wave at non-normal incidence becomes an even more difficult task [13–15].

All sources implemented in our simulations are fully described elsewhere [13]. Here, we limit ourselves to say where and how the different sources are useful.

i. *Gaussian wave-packet* A gaussian wave-packet is a good source for illuminating 1D and 2D periodic systems at normal incidence. It has the advantage to be compact in space and broadband in the frequency domain. This source is settled at $t = 0$.
   Normal incidence is definedas the direction perpendicular to the film where the lattice is defined. In our notation this direction coincides with the z-direction. One dimensional periodic systems can be considered as a particular case of the 2D-periodic case, where the system is invariant along one of the in-plane directions. Furthermore, at normal incidence, $E_z = H_z = 0$, so:

$$\vec{E}_{||}(x, y, z, t) = \vec{E}^o_{||} e^{\imath \frac{\omega_o}{c}(z - z_o - ct/n)} e^{-\left(\frac{z - z_o - ct/n}{D}\right)^2} \tag{1.27}$$

   Here z_o is the position where the initial gaussian field reaches its maximum value, where $\omega_o = \frac{2\pi c}{\lambda_o}$ and n the refractive index. The initial magnetic field is obtained from Maxwell's equations.
   If we define the Fourier's transformation as $f(\omega) = \int_{-\infty}^{+\infty} dt f(t) e^{\imath \omega t}$ we find that

$$\vec{E}_{||}(x, y, z, \omega) = \frac{\sqrt{\pi} n D}{c} \vec{E}^{o}_{||} e^{\imath \frac{n\omega}{c}(z - z_o)} e^{-\left(\frac{nD}{2c}\right)^2 (\omega - \omega_o)^2} \tag{1.28}$$

"Playing" with the D parameter it is possible to fit the gaussian wave-packet to the available computational space. Note that in the limit of $D \to \infty$ (Eq. 1.27) the wave-packet becomes a monochromatic plane wave.

ii. *Dipole sources* Dipole sources in FDTD are useful for calculating dispersion relations. In these situations we want a source able to couple with all the EM modes of a given structure, which we do not know beforehand. A dipole source can be tuned to be broadband or monochromatic. Moreover, all the k-vectors can be accessed with a dipole source. Dipole sources can be settled to mimic either a magnetic or an electric dipole, so with such a source we can take advantage of system symmetries. We will use three types of dipoles:

$$\begin{aligned} &\delta(\vec{r} - \vec{r}_o) \times \delta(t - t_o) \\ &\delta(\vec{r} - \vec{r}_o) \times e^{-\imath \omega_0 t} \\ &\delta(\vec{r} - \vec{r}_o) \times e^{-\imath \omega_0 t} e^{-\left(\frac{t - t_0}{\tau}\right)^2} \end{aligned} \tag{1.29}$$

The first type is broadband in frequency (and is switched on at $t = 0$). The second and the third types must be updated in time. The second type represents a monochromatic source while the third one is broadband in frequency. These sources emit both propagating and evanescent waves, thereby are useful in order to "probe" confined modes, unaccessible for a propagating wave.

iii. *Sum of Bloch's waves* In periodic systems the EM modes are a superposition of Bloch's waves. The best way to access them is precisely by an illumination with a superposition of such waves. This source was first used in FDTD by Chan et al. [16]. Again we refer to Taflove's book [13] for a complete description.

1.2.3 Data Processing

Calculation of Optical Spectra: "Projection of EM Fields onto Plane Waves"

Maxwell's equations are solved in real space and in time domain with FDTD, in other words, a single FDTD simulation results in the knowledge of $\vec{E}(\vec{r}, t)$ and $\vec{H}(\vec{r}, t)$. Nevertheless, these vectors do not provide the most relevant information about the optical properties by themselves. Actually, the optical response of a certain structure is described in terms of scattering coefficients, transmission/reflection spectra, near field maps, dispersion relations... The optical response usually depends on the pumping frequency (even though the materials involved are non-dispersive). To obtain a frequency dependent quantity is mandatory to apply a Fourier's transformation to the EM fields in the time domain,

$$\vec{E}(\vec{r},\omega)=\frac{1}{\sqrt{2\pi}}\int_{-\infty}^{+\infty}\vec{E}(\vec{r},t)e^{-\imath\omega t}\,dt$$

$$\vec{H}(\vec{r},\omega)=\frac{1}{\sqrt{2\pi}}\int_{-\infty}^{+\infty}\vec{H}(\vec{r},t)e^{-\imath\omega t}\,dt \tag{1.30}$$

which is not always straightforward as we will see.

Let us concentrate first on how transmitted and reflected energy currents from a material layer can be calculated with FDTD. In fact, these quantities are not difficult to calculate, once the EM fields ($\vec{E}(\vec{r},\omega)$ and $\vec{H}(\vec{r},\omega)$) are known. The averaged Poynting vector flowing through a given surface, S, reads (in the MKS system of units),

$$J(\omega)=\frac{1}{2}\int_{S} d\vec{s}\,Re\left[\vec{E}(\vec{r},\omega)\times\vec{H}^{*}(\vec{r},\omega)\right] \tag{1.31}$$

Equation 1.31 provides the total energy current, at a fixed frequency. In order to obtain transmission and reflection spectra we would need to know the incident EM fields. For this, it is necessary to run an extra simulation without scatterers if the incoming fields are not analytically known. But here, we will discuss a different method to calculate the transmission and reflection coefficients in periodic systems, which improves to some extent the previous one. As compared with it, this method is interesting for two additional reasons. It is possible to obtain transmission and reflection coefficients which contain information both in the frequency domain and in the reciprocal space. Moreover, this method allows also to calculate separately transmission and reflection from a single simulation. The basic idea consists in finding a way to isolate the current that each $\vec{k}$-vector of the reciprocal lattice carries, as a function of both the wavelength and the polarization state (See Ref. [17] for further details).

The plane wave solution for Maxwell's equations have the following form:

$$\vec{E}(\vec{r},t)=\vec{E}_0 e^{\imath(\vec{k}\vec{r}-\omega t)},\ \vec{H}(\vec{r},t)=\vec{H}_0 e^{\imath(\vec{k}\vec{r}-\omega t)} \tag{1.32}$$

where $\omega=\frac{c|k|}{\sqrt{\varepsilon}}$, being the speed of light in vacuum, c, and ε the dielectric constant of such media. Thus, curl Maxwell's equations (MKS system of units) can be written as:

$$\vec{k}\times\vec{E}_0=\mu_0\omega\vec{H}_0$$

$$\vec{k}\times\vec{H}_0=-\varepsilon\varepsilon_0\omega\vec{E}_0 \tag{1.33}$$

For a given $\vec{k}$ vector there are two polarization states that must be considered, because Eq. 1.33 are invariant under simultaneously change $\vec{E}\rightarrow-\mu_0\vec{H}$ and

$\vec{H} \rightarrow \varepsilon\varepsilon_0 \vec{E}$. We use the usual notation for such states, that is, s-polarized plane waves are defined as,

$$\vec{E}_s \propto \vec{k} \times \vec{n} \tag{1.34}$$

and for the p-polarization,

$$\vec{E}_p \propto \vec{k} \times \vec{k} \times \vec{n} \tag{1.35}$$

where $\vec{n}$ is an arbitrary unit vector. The propagation $\vec{k}$ vector and $\vec{n}$ define a plane in space with respect to, the electric (magnetic) field oscillates perpendicular for the s-polarization (p-polarization).

We restrict our analysis to 2D periodically structured systems, and to the transmission and reflection coefficients in the far field. The 2D-lattice would define the $x-y$ plane, and $\vec{n}$ would be the unit vector $\hat{z}$.

At fixed frequency, a plane wave is completely described by the components of the wave vector parallel to the surface ($\vec{k}_{||}$) and its polarization. We use σ for labeling the polarization state, which can be either $+p$, $-p$, $+s$ or $-s$. The sign accounts for the direction the plane wave propagates, i.e, as $k_z = \pm\sqrt{\varepsilon(\frac{\omega}{c})^2 - \vec{k}_{||}^2}$, the $\pm$ signs denotes the plane waves propagating coming from $\mp\infty$, respectively. Evanescent waves ($k_z = \imath|k_z|$) do not carry energy to the far field, so they will not be considered in the following.

Moreover, we are interested only in the EM field components parallel to the $x-y$ plane, which contain the necessary information to compute the time averaged Poynting's vector flow,

$$\langle S_z \rangle = E_x H_y^* - E_y H_x^* \tag{1.36}$$

As we have set $\vec{n} = \hat{z} = (0, 0, 1)$, taking into account Eq. 1.35 and Maxwell's equations we find on one hand,

$$\vec{E}^0_{\vec{k}_{||},p} = \beta k_z \left(k_x, k_y, -\frac{(k_x^2 + k_y^2)}{k_z} \right)^T$$

and

$$\vec{H}^0_{\vec{k}_{||},p} = \frac{\beta}{\mu_0 \omega} |k|^2 \left(-k_y, k_x, 0\right)^T \tag{1.37}$$

where T stands for transposition. On the other hand, by utilizing Eq. 1.34 it can be shown that:

$$\vec{E}^0_{\vec{k}_{||},s} = \beta \left(-k_y, k_x, 0\right)^T$$

and

$$\vec{H}^{0}_{\vec{k}_{||},p} = \frac{\beta}{\mu_0 \omega} k_z \left(k_x, k_y, -\frac{\left(k_x^2 + k_y^2\right)}{k_z} \right)^T \quad (1.38)$$

The value for β can be arbitrarily chosen, however it is usually chosen so that the current carried by the wave is the unity. A complete description of an eigenvector in "free" space, at fixed frequency, can be expressed in this way:

$$\langle \vec{r} | \vec{k}_{||}, \sigma \rangle = \vec{f}^{\,r}_{\vec{k}_{||},\sigma} e^{\imath \vec{k}\vec{r}} \quad (1.39)$$

where we have used Dirac's notation, and $\vec{f}^{\,r}_{\vec{k}_{||},\sigma}$ denotes the different polarization state of the right-vectors:

$$\vec{f}^{\,r}_{\vec{k}_{||},\sigma} = \left(e_x, e_y, h_x, h_y\right)^T \quad (1.40)$$

In the last expression the field components are those shown in the set of Eqs. 1.37 and 1.38. Note that for the particular case where $\vec{k}_{||} = \vec{0}$ we must choose the basis element by hand. We choose therefore at a fixed wavelength, a bi-vector EM field, $\vec{F}(\vec{r}, \omega)$, can be described as,

$$|F\rangle = \int d\vec{k}_{||} \sum_{\sigma} \alpha(\vec{k}_{||}, \sigma) |\vec{k}_{||}, \sigma\rangle \quad (1.41)$$

In each time step, the FDTD method output is precisely the EM field at this loop iteration. Within this framework, the EM field components at certain z_0 can be written as:

$$\vec{F}(\vec{r}_{||}, z_0, t) = \int d\omega \vec{F}(\vec{r}_{||}, z_0, \omega) e^{-\imath\omega t} \quad (1.42)$$

where

$$\langle \vec{r} | F \rangle = \vec{F}(\vec{r}_{||}, z_0, \omega) = \sum_{\sigma} \int d\vec{k}_{||} \alpha(\vec{k}_{||}, \omega, \sigma) \langle \vec{r} | \vec{k}_{||}, \sigma \rangle \quad (1.43)$$

To obtain $\alpha(\vec{k}_{||}, \omega, \sigma)$ we must project $|F\rangle$ onto the left-vector basis $\{\langle \vec{k}_{||}, \sigma |\}$. Unfortunately, the right eigenvectors do not in general form an orthonormal set, so the left ones must be found by inverting the matrix built with the right-vectors [17].

In fact, the FDTD method has a great advantage over others: a single simulation is enough to provide the optical response as a function of frequency. However, Fourier's integral calculations are time consuming processes. To avoid this drawback as much as possible, one can make use of the fast fourier transformation (FFT). Usually, the FFT method is the best choice in post-processing. However the use of FFT methods to evaluate (1.30) or (1.42) requires storing the fields in the computer memory for all times, which is usually prohibitive. Alternatively, if the Fourier's integral is done by adding the contributions for each "time slice" as time evolves, fields do not need to be stored, but performing the fourier transform (FT) is computationally costly. Therefore the best choice, depends on the problem we are studying and on the computer resources (speed and available RAM memory).

For the type of structures investigated in next chapters, storing the EM fields of a typical simulation at each time step is a hard constrain. When the system under study is large, i.e., when the computer RAM memory requirements are too demanding, the FFT is not a feasible approach. From a single FDTD simulation the left-hand side of Eq. 1.42 is obtained, leaving the calculation of $\alpha(\vec{k}_{||}, \omega, \sigma)$ to the FFT post-processing. Let us show how this way to proceed can not be followed for simulating transmission or reflection spectra through 2D-systems. Typically, we investigate 2D-periodic structures with periods ranging from 300 nm to 1000 nm. As we said before, the mesh size must be quite fine to ensure accuracy (5–10 nm). The film where the array is patterned is usually 25–500 nm thick. Overall, the whole system (including, vacuum, PMLs,...) is about $X \times Y \times Z = 100 \times 100 \times 300$ mesh points. Note each point at this grid would imply to store six complex numbers (EM components) plus certain auxiliary variables. A system like that would require well over 2Gb RAM. For instance, to compute $\alpha(\vec{k}_{||}, \omega, \sigma)$ on a single layer of constant z would mean storing a slice 100×100, one for each time step. Converged results are typically obtained within the range from 30,000 to 120,000 time steps, so it would needed an available memory from 200 to 400 Gb!. It is obvious that FFT can not be used for these systems. The best way to proceed in this case consists on calculating the Fourier's integral directly:

$$\alpha(\vec{k}_{||}, \omega, \sigma) = \int dt \left[\int d\vec{r}_{||} \langle \vec{k}_{||}, \sigma | \vec{r} \rangle z \vec{F}(\vec{r}_{||}, z_0, t) \right] e^{\imath \omega t} \tag{1.44}$$

Note that, fixed the frequency, ω, each pair $(\vec{k}_{||}, \sigma)$ defines an element of a basis in which an arbitrary EM field can be expanded. At this point, we have an infinite number of eigenvector coefficients to be calculated. However, in 2D periodic systems, only a finite number of such elements carry energy to the far field. First, Bloch's theorem [9] imposes that only the reciprocal lattice vectors contribute to the integral [18], thus:

$$|F\rangle = \sum_{\vec{G}} \sum_{\sigma} \alpha(\vec{G}, \sigma) |\vec{G}, \sigma\rangle \tag{1.45}$$

where $\vec{G}$ runs over the reciprocal lattice vectors defined as $\vec{G}\vec{R} = \delta_{ij}$, for all lattice vectors $\vec{R}$. Up to here, we have reduced the number of integrals to calculate, although we still have an infinite numerable number of them. Luckily only a finite number of these coefficients represent propagative $\vec{k}$ vectors (for which k_z is a real number). Therefore, in most of the calculations only a few coefficients in (1.44) must be calculated in order to find transmission and reflection currents.

Band Structure and Dispersion Relation Calculations

Band structure and dispersion relation curves provide fundamental information about the EM modes supported by a given structure. Next, we will discuss how calculate them with the FDTD method.

It is not difficult to implement an algorithm in order to calculate band structures for periodic systems with FDTD. Fortunately, a periodic system can be represented by a single unit cell within FDTD. Bloch's boundary conditions supply with the interactions between neighbor cells, thus providing the optical response as a function of the incident k-vector along a chosen periodic direction. The source of illumination used to be a sum of Bloch's waves, though dipole sources (Sect. 1.2.2) work as well. The key point is that the source fields should somehow match the EM modes sustained by the structure. Fixed the boundary conditions (k-vector) and the field source, maxima in spectra (calculated at "suitable" points in space) settle the EM modes of the structure. Repeating the last procedure for each wave-vector belonging the first Brillouin zone, it can be finally obtained the band structure. Indeed, the FDTD band structures calculated throughout this thesis were calculated using this technique.

On the other hand, if the structure is invariant through a given direction in space, the dispersion relation can be straightforwardly found using the same method. Suppose that we are interested on the dispersion relation through that direction, which is denoted as z, so $k_z = k_z(\omega)$ represents the wave-vector through it as a function of the frequency, ω. The whole system can be then fitted in a single "slice", containing its profile, which repeats itself along z. The slice plays the role of the unit cell of a periodic system for which the period coincides with the mesh size. The initial problem of calculating a dispersion relation is "mapped" onto a more easy problem, i.e., to calculate a band structure within the first Brillouin zone, $|k_z| \leq \pi/q$, where q is the mesh size. Note that the smaller the mesh size, the longer the first Brillouin zone in k-space is. Because of q is usually very small as compared to the wavelength (to ensure convergency, accuracy,...), the first Brillouin zone so defined spreads over a wide range of k-vectors without being folded onto it. As an example, we show in Fig. 1.6 the FDTD calculated dispersion relation of SPPs supported by a semi-infinite gold film (circular symbols). This 1D problem is one of the simplest that can be treated with FDTD. The system is divided in two different half spaces (metal and vacuum). A dipole just over the metal surface is chosen to be the EM field source, so that its evanescent fields overlap with SPPs. The "probe", at which the field amplitude of either the magnetic or electric field is calculated, is positioned a few nanometers inside the metal. As we see, there is good agreement between FDTD and the analytical SPP dispersion relation (solid line). In this case case, the first Brillouin zone extents as far as 630 μm^{-1} (the mesh size used is 5 nm), wide enough to cover the frequency range of interest.

In general, the k-vector is a complex number ($k = k_r + \imath k_i$). Up to here we have been concerned with the real part of the dispersion relation, i.e., $k_r = k_r(\omega)$. The imaginary part (k_i) defines the propagation length. Hereafter, we define it as the distance at which the field intensity has decayed a $1/e$ factor, so $l_{\text{abs}} = [2k_i]^{-1}$. As we

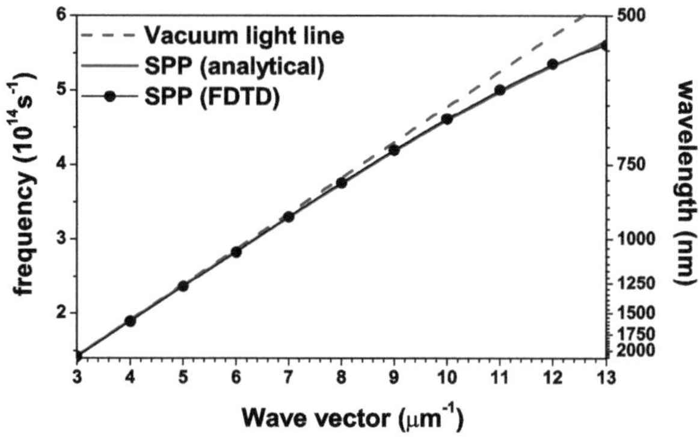

Fig. 1.6 SPP dispersion relation for gold: analytical (*solid curve*) and calculated with the FDTD method (*circular symbols*). The mesh size used is 5 nm. The *dashed line* depicts the light cone

will see in next chapters, propagation length is one of the most important properties of guided EM modes at the nanoscale. In fact, one could calculate propagation lengths running 3D-FDTD simulations. Illuminating the system (e.g. with a dipole) at a given "point" and then picking the field up at several relevant points, the propagation length could be directly obtained from fields in real space. The last would require huge systems, and even the problem of how the structure would be illuminated is difficult to solve. We have chosen another way to proceed that allow us extract both the real and imaginary part of k running a single simulation. We assume that EM fields are harmonic in time, thus $\phi(t) \propto e^{-\imath\omega_r t}e^{-\omega_i t}$, where ω_i must be chosen positive so that the fields exponentially decay. Additionally, let us express $\phi(t)$ in the frequency domain, that is, $\phi(\omega) \propto \int dte^{-\imath(\omega-\omega_r)t}e^{-\omega_i t} \propto \frac{1}{(\omega-\omega_r)+\imath\omega_i}$. Therefore:

$$|\phi(\omega)|^2 \propto \frac{1}{(\omega-\omega_r)^2+\omega_i^2} \tag{1.46}$$

thus $2\omega_i = \text{FWHM} = \Delta\omega$, where FWHM states for the acronym of full-width at half-maximum. In this case, $k_i = \frac{\omega_i}{v_g} = \frac{\Delta\omega}{2\,v_g}$ (v_g being the group velocity), so finally:

$$l_{\text{abs}} = \frac{v_g}{\Delta\omega} \tag{1.47}$$

In summary, because of time harmonic response of EM fields, we are "probing" not only the location of the spectral positions at the $k_r(\omega)$ plane with this method, but also the propagation length, retrieved from the FWHM of the spectrum resonances.

We turn to SPP properties. Figure 1.7 renders l_{SPP} analytical values (solid line) compared to those calculated with Eq. 1.47 by means of the FDTD method (symbols). Different curves render different sizes of the vacuum half space (N_z), for different simulation times (T_{max}). Interestingly, the calculated propagation lengths are smaller than the analytical ones at large wavelengths, all except for the case in which $N_z = 50\,\mu$m and $T_{\text{max}} \sim 30{,}800$ fs. The explanation of this behavior is quite simple: at large wavelengths SPPs get less absorbed within the metal, furthermore they are less confined in vacuum, so SPPs can stand on the surface for a long time until the

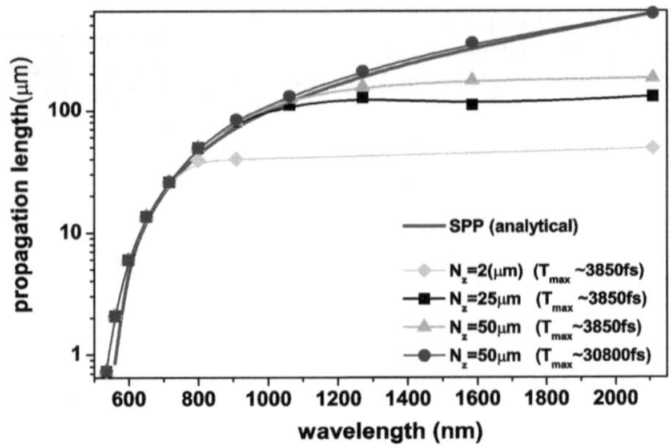

Fig. 1.7 SPP propagation length for gold: analytical (*continuous curve*) and calculated with the FDTD method (*symbols*). The distance from the metal surface to the corresponding PML is denoted by N_z, and T_{max} renders the total time that a simulation takes

resonance builds up, spreading a lot far from the surface. Note that our method relies on being able to accurately calculate the FWHM from the spectral response, and precisely this magnitude strongly depends on the time the SPP stands on the surface. This explains that so long time consuming simulations were needed to get good results. On the other hand, if the space between PMLs and the metal surface is smaller than a SPP skin-depth in vacuum, SPP may be absorbed by the PMLs, broadening the resonance and thereby the FWHM too. This explain the slightly improvement shown as the vacuum region expands from $N_z = 2\,\mu$m to $N_z = 25\,\mu$m. Therefore, in order to calculate l_{abs} for an EM mode that propagates through a given structure, one must carefully choose the simulation time. Besides, space regions surrounding the system must be allocated in the FDTD mesh ensuring they are large enough to fit it.

1.2.4 Metals Within the FDTD Approach

The Perfect Electric Conductor Approximation

A very useful approximation to investigate the EM properties of metals consists on considering them as perfect electric conductors (PECs). Roughly speaking, the PEC approximation disregards the penetration of the EM fields into the metal. The latter considers the metal conductivity as infinite, i.e., charges inside the metal instantaneously respond to the optical excitation. The PEC approach is a very good approximation for metals at microwave or terahertz frequencies. At optical frequencies the PEC approximation misses some important phenomena (as the existence of SPPs). Nevertheless, even at optical frequencies the PEC approximation is quite often an useful starting point for the theoretical analysis.

Apparently, this approximation could seem easy to implement within a FDTD scheme: at the metal surface the electric field component parallel to it must be set to

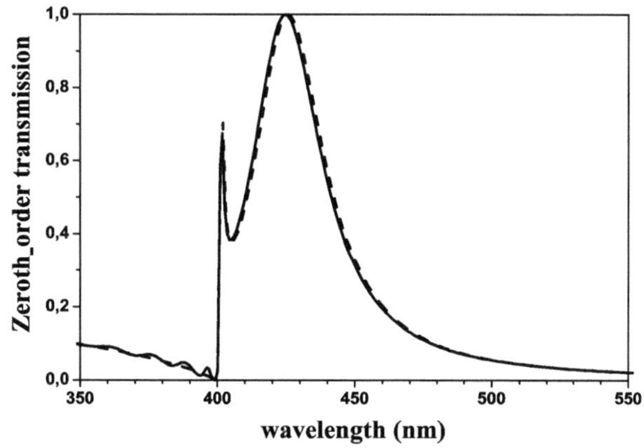

Fig. 1.8 Zeroth-order transmission through a 2DHA of *square holes* (side = 200 nm), perforated through a free-standing metal film (width = 100 nm) approximated as it were a PEC. The period is 400 nm. The *solid line* depicts the calculation with the FDTD method. The *dashed line* is obtained with the CMM (Sect. 1.3)

zero. However, the EM field distribution induced by the Yee's cell requires the implementation of this boundary condition to be handle with care. Consider a structure where the PEC regions are in contact with other materials, dielectrics for instance. In the continuous space, frontiers between both media are well defined. However, when the continuous space is divided in small cubes (like in the FDTD algorithm), we must fix them by hand. Let us explain this more precisely. The FDTD algorithm operates on a discretized space, where the whole space is filled by Yee's cubes. The faces of such cubes provide us with suitable boundaries. This implies that some cells have some of their EM components "on" the boundary while others are only close to. This is, there are no "metal" cells and "dielectric" cells. Instead, the PEC boundaries must be defined by Yee's cell faces. Once this is clear the implementation of the PEC approximation on the FDTD code is a question of careful identification of those Yee's faces that require special treatment, for any given metal structure. We have implemented the PEC approximation on the "home-made" FDTD code used in this thesis and in order to show that it works, we compare in Fig. 1.8. transmission spectra through a two-dimensional hole array (2DHA) of square holes in a PEC film calculated with two different techniques: FDTD and the coupled mode method (CMM) (See Sect. 1.3). For PEC metals CMM is exact, and as we can observe FDTD recovers the exact result.

Dispersive Materials

Dispersive materials require a special treatment in FDTD, as the dielectric constant is local in the frequency domain but non-local in the time domain. The Maxwell's equations for i.h.l media in the MKS system of units are

$$\nabla \times \vec{E}(\vec{r},t) = -\mu_0 \frac{\partial \vec{H}(\vec{r},t)}{\partial t} \tag{1.48}$$

$$\nabla \times \vec{H}(\vec{r}, t) = \frac{\partial \vec{D}(\vec{r}, t)}{\partial t} \tag{1.49}$$

Non-locality in time-domain generally implies that $\vec{D}(\vec{r}, t) \neq \alpha \vec{E}(\vec{r}, t)$, where α is a constant. However, in the frequency domain the electric field and the displacement vector are proportional

$$\vec{D}(\vec{r}, \omega) = \varepsilon(\omega) \vec{E}(\vec{r}, \omega) \tag{1.50}$$

where the dielectric constant ε, links $\vec{E}$ and $\vec{D}$, at fixed frequency ω.

Remember the FDTD algorithm operates in the time domain, so when Maxwell's equations are discretized we should count on a time domain version of (1.50), i.e., its convolution

$$\vec{D}(t) = \varepsilon_0 \varepsilon \vec{E}(t) + \varepsilon_0 \int_{\tau=0}^{t} \vec{E}(t - \tau) \chi(\tau) d\tau \tag{1.51}$$

where $\chi(\tau)$ is the first order electric susceptibility in the time domain (From now on the explicit dependence in the space coordinates will be omitted.)

Throughout this work we have used one of the methods for incorporating dispersive properties available in FDTD [13]. The first FDTD approach for simulating realistic dispersive materials was conducted by Luebbers et al. [19]. They started investigating substances with an optical response well described by the Debye model. Next, they extended their conclusions to metals behaving like plasmas [20]. Finally, they took also into account effects due to the interband transition of electrons in metals [21]. They called this general procedure piece linear recursive convolution (PLRC) method.

In the course of this thesis we have been mainly interested in how light interacts with nano-structured metals, at wavelengths ranging from the visible regime to the Terahertz regime. Metals at those frequencies are well described by the Drude–Lorentz model, where the dielectric constant is fitted by several Drude-like and Lorentzian terms:

$$\epsilon(\omega) = \epsilon_r - \sum_j \frac{\omega_{pj}^2}{\omega(\omega + \imath\gamma_j)} - \sum_j \frac{\Delta\epsilon_j \Omega_j^2}{\omega^2 - \Omega_j^2 + \imath\omega\Gamma_j}. \tag{1.52}$$

In fact, in the PLRC method an arbitrary number of Drude–Lorentz terms can be straightforwardly incorporated, which constitutes the main advantage of this method. Let us briefly outline the PLRC method. In the PLRC method a discretized version of the integral appearing in (1.51) is considered:

$$I(n\delta t) = \sum_{m=0}^{n-1} \int_{m\delta t}^{(m+1)\delta t} \vec{E}(n\delta t - \tau) \chi(\tau) d\tau \tag{1.53}$$

A very good approximation for the electric field value at time t is obtained by choosing $\vec{E}(t) = \vec{E}^i + \frac{[\vec{E}^{i+1} - \vec{E}^i]}{\delta t}(t - i\delta t)$, where $\vec{E}^i = \vec{E}(i\delta t)$. In other words, the electric field is approximated by truncating the corresponding Taylor's series up to the linear term

$$I(n\delta t) = \sum_{m=0}^{n-1} \vec{E}^{n-m} \int_{m\delta t}^{(m+1)\delta t} \chi(\tau)d\tau + \sum_{m=0}^{n-1} \frac{[\vec{E}^{n-m-1} - \vec{E}^{n-m}]}{\delta t} \int_{m\delta t}^{(m+1)\delta t} (m\delta t - \tau)\chi(\tau)d\tau \tag{1.54}$$

It is convenient to define

$$\chi_m = \int_{m\delta t}^{(m+1)\delta t} \chi(\tau)d\tau$$
$$\xi_m = \frac{1}{\delta t} \int_{m\delta t}^{(m+1)\delta t} (m\delta t - \tau)\chi(\tau)d\tau \tag{1.55}$$

Therefore,

$$\vec{D}^n = \varepsilon_0 \varepsilon \vec{E}^n + \varepsilon_0 \sum_{m=0}^{n-1} \left\{ \chi_m \vec{E}^{n-m} + \xi_m [\vec{E}^{n-m-1} - \vec{E}^{n-m}] \right\} \tag{1.56}$$

As we know the right hand side of Eq. 1.49 is approximated within the FDTD formalism as $(\vec{D}^{n+1} - \vec{D}^n)/\delta t$, but evaluated at the same spatial position. In this way the last equation can be readily written as follows

$$\vec{D}^{n+1} - \vec{D}^n = \varepsilon_0(\varepsilon + \chi_0 - \xi_0)\vec{E}^{n+1} + \varepsilon_0(\xi_0 - \varepsilon)\vec{E}^n - \varepsilon_0 \vec{\Psi}^n \tag{1.57}$$

where

$$\vec{\Psi}^n = \sum_{m=0}^{n-1} \left[\Delta\chi_m \vec{E}^{n-m} + \Delta\xi_m (\vec{E}^{n-m-1} - \vec{E}^{n-m}) \right] \tag{1.58}$$
$$\Delta\chi_m = \chi_m - \chi_{m+1}$$
$$\Delta\xi_m = \xi_m - \xi_{m+1}$$

Furthermore, if we assume different contributions to the electric susceptibility in the form of $\chi(\omega) = \sum_{i=1}^{p} \chi^i(\omega)$, in the time domain we will have $\chi(t) = \sum_{i=1}^{p} \chi^i(t)$. Finally, the expression for updating $\vec{E}(\vec{r}, t)$ reads,

$$\vec{E}^{n+1} = \frac{\varepsilon - \xi_0}{\varepsilon - \chi_0 + \xi_0}\vec{E}^n + \frac{\delta t/\varepsilon_0}{\varepsilon - \chi_0 + \xi_0}\nabla \times \vec{H}^{n+1/2} + \frac{1}{\varepsilon - \chi_0 + \xi_0}\sum_{i=1}^{p}\vec{\Psi}_i^n \quad (1.59)$$

To implement a FDTD algorithm for dispersive materials, one should start taking into account these expressions. However, a last ingredient is needed to use them: the time domain response $\chi(t)$ for a given $\epsilon(\omega)$. For a Drude term the inverse Fourier's transformation ($\mathfrak{F}^{-1}$) of $\varepsilon(\omega) - \varepsilon_r$ yields

$$\mathfrak{F}^{-1}\left(-\frac{\omega_p^2}{\omega(\omega + \imath\gamma)}\right) \Rightarrow \frac{\omega_p^2}{\gamma}\left(1 - e^{-\gamma t}\right)U(t) \quad (1.60)$$

where $U(t)$ is the time-step function.

On the other hand, a Lorentz term results in

$$\mathfrak{F}^{-1}\left(-\frac{\Delta\epsilon\Omega^2}{\omega^2 - \Omega^2 + \imath\omega\Gamma}\right) \Rightarrow \delta e^{-\alpha t} sin(\beta t)U(t) \quad (1.61)$$

where

$$\alpha = \frac{\Gamma}{2}$$
$$\beta = \sqrt{\Omega^2 - \alpha^2}$$
$$\delta = \frac{\Delta\varepsilon\Omega^2}{\beta}$$

Luebbers et al. realized that $\vec{\Psi}^n$ could be recursively calculated if $\chi(t)$ is represented as an exponential-in-time function. Otherwise, it would be necessary to store $\vec{E}^n$ for a large number of previous time slices, which would be very inefficient. Therefore, the recursive updating of $\vec{\Psi}^n$ imposes a hard constraint: only specific functional dependencies of the dielectric constant are suited within the PLRC algorithm, as we will show next.

Each term $\vec{\Psi}^n$ in Eq. 1.59 can be expressed as two sums, and each one can in turn, be represented as follows

$$\psi^n = \sum_{m=0}^{n-1} f^{n-m}\alpha^m \quad (1.62)$$

where we have disregarded the vector notation for simplicity. Here $f^{n-m}\alpha^m$ is either $\vec{E}^{n-m}\Delta\chi_m$ or $(\vec{E}^{n-m-1} - \vec{E}^{n-m})\Delta\xi_m$ in Eq. 1.59.

The updating of ψ^n is simplified if $\alpha^{m+1} = k\alpha^m$ (being k a constant). In the case of Drude terms, it is easy to demonstrate from Eq. 1.60 that both $\Delta\chi_m$ and $\Delta\xi_m$ have such a form. Then, ψ^{n+1} can be evaluated as:

$$\psi^{n+1} = \sum_{m=0}^{n} f^{n+1-m}\alpha^m = \alpha^0 f^{n+1} + \sum_{m=1}^{n} f^{n+1-m}\alpha^m \quad (1.63)$$

taking $j = m - 1$ we find

$$\psi^{n+1} = \alpha^0 f^{n+1} + \sum_{j=0}^{n-1} f^{n-j} \alpha^{j+1} \tag{1.64}$$

as $\alpha^{j+1} = k\alpha^j$, we reach this final result

$$\psi^{n+1} = \alpha^0 f^{n+1} + k \sum_{j=0}^{n-1} f^{n-j} \alpha^j = \alpha^0 f^{n+1} + k\psi^n \tag{1.65}$$

However, due to the form we express the Lorentz time dependence $\chi(t)$ (See Eq. 1.61), we can not recursively update the corresponding discrete convolution. Fortunately, we can define a complex time-domain susceptibility like this

$$\hat{\chi}(t) = -\imath\delta e^{-(\alpha+\imath\beta)t} U(t) \tag{1.66}$$

so that $\chi(t) = Re[\hat{\chi}(t)]$.

In the last case, if the EM fields are real numbers, it is straightforward to demonstrate that the equation for updating ψ^n becomes,

$$\psi^{n+1} = Re[\hat{\alpha}^0] f^{n+1} + k \sum_{j=0}^{n-1} f^{n-j} Re[\hat{\alpha}^j] = Re[\hat{\alpha}^0] f^{n+1} + k\,\psi^n. \tag{1.67}$$

What does it happen in the case of complex fields? Note that in that case:

$$\psi^n = \sum_{m=0}^{n-1} Re[f^{n-m}]\alpha^m + \imath \sum_{m=0}^{n-1} Im[f^{n-m}]\alpha^m \tag{1.68}$$

Provided α^m is proportional to an exponential function (like in the Drude case) the term involving $Im[f^{n-m}]$ does not add new difficulties. Nevertheless, Lorentz contributions need to hold a previous condition, that is α^m must be expressed as a complex number. In order to find the updating formula we start with the complex version of ψ^n

$$\hat{\psi}^n = \sum_{m=0}^{n-1} Re[f^{n-m}]\hat{\alpha}^m + \imath \sum_{m=0}^{n-1} Im[f^{n-m}]\hat{\alpha}^m \tag{1.69}$$

or $\hat{\psi}^n = \hat{\psi}^n_R + \imath\,\hat{\psi}^n_I$, so $\psi^n = Re[\hat{\psi}^n_R] + \imath\, Re[\hat{\psi}^n_I]$. This is the correct updating expression when dealing with realistic metals and complex EM fields in the FDTD algorithm. To our knowledge, a PLRC algorithm has not been reported so far for properly updating EM complex fields in the general Drude–Lorentz approximation of the dielectric constant of metals. To illustrate how advantageous it is, we have

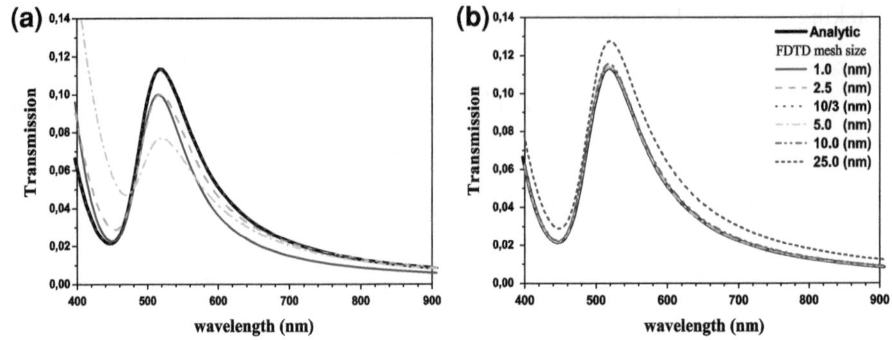

Fig. 1.9 FDTD calculated transmission spectra at normal incidence through a 50 nm thick gold film, for different mesh sizes: **a** without and **b** with the PLRC correction for updating complex fields

plotted an example in Fig. 1.9. This figure shows FDTD calculations of transmission spectra at normal incidence through a 50 nm thick gold film for different mesh sizes. The dielectric constant of gold has been taken from the experimental data and fitted to a Drude–Lorentz formula (Table 1.1). If the "complex updating" is not done (Fig. 1.9a), the EM fields are updated as in the literature [21], and what it is obtained is disconcerting. The expected trend in FDTD does not hold, i.e, the finer the mesh size, the worse the results. However, the results depicted in panel(b) are correct, where the "complex" updating procedure (Eq. 1.69) has been taken into account. In the last case, the transmission curves calculated with the FDTD method converge to the analytical one, with mesh sizes smaller than 10 nm. As we said at the introductory subsection, the FDTD accuracy in treating metals depends so much on how the EM fields are sampled inside the metal. Even so, a relatively good agreement is achieved even for a 25 nm mesh size, albeit the skin-depth (δ_m) is of the same order.

1.2.5 Outer Boundary Conditions

Still, the FDTD method described up to now would have very limited applicability. The reason is that most of the interesting phenomena have a resonant nature, which implies that the optical response of the structure builds up over a long period of time. Not only the time evolution of fields at the structure must be followed, but also the fields that have been radiated away in previous times. A sloppy treatment of these fields would introduce spurious fields back into the system. For instance, setting them to zero at some distant region from the system of interest would be equivalent to placing a perfectly reflecting mirror, not to the disappearance of these fields. The ideal would be to terminate the system with a sort of "material" able to perfectly absorb light, but without being reflected when light impinges on it. In FDTD there are several ways to achive this. The most extended and used method is based on the

Table 1.1 Dielectric constant fitted parameters of Ag, Au, Cu, Al, Ni, Cr and W. ω_P, γ, Γ and Ω in electron volts

	Ag	Au	Cu	Al	Ni	Cr	W
ϵ_r	4.6	5.967	1.0	1.0	1.0	1.0	1.0
ω_{P0}	9.0	8.729	8.212	10.83	4.621	4.406	5.955
γ_0	0.07	0.065	0.03	0.047	0.021	0.047	0.027
ω_{P1}	-	-	-	-	6.929	-	2.286
γ_1	-	-	-	-	1.771	-	0.335
ω_{P2}	-	-	-	-	7.062	-	-
γ_2	-	-	-	-	3.443	-	-
Γ_0	1.2	0.433	0.378	0.333	1.021	3.175	0.590
Ω_0	4.9	2.684	0.291	0.162	1.458	0.121	0.984
$\Delta\epsilon_0$	1.10	1.09	84.49	1940.97	2.1	1191.85	12.0
Γ_1	-	-	1.056	0.312	2.410	1.305	1.653
Ω_1	-	-	2.957	1.544	3.443	0.543	2.066
$\Delta\epsilon_1$	-	-	1.395	4.706	1.2	58.79	14.4
Γ_2	-	-	3.213	1.351	-	2.676	2.479
Ω_2	-	-	5.3	1.808	-	1.970	4.132
$\Delta\epsilon_2$	-	-	3.018	11.39	-	34.21	12.9
Γ_3	-	-	4.305	3.382	-	1.335	-
Ω_3	-	-	11.18	3.473	-	8.775	-
$\Delta\epsilon_3$	-	-	0.598	0.558	-	1.238	-

so-called perfect-matched-layers (PMLs) first proposed by Berenguer [22]. We omit here the details that are fully developed in Ref. [13], and references therein. Since the first algorithm by Berenguer, the PML boundary conditions have been improved a lot. In fact, we use in our simulations Uniaxial PMLs (UPMLs). Roughly speaking, these boundary conditions are equivalent to a uniaxial and dispersive material with the above-mentioned optical properties. However, these absorbing layers suffer from an important drawback, the absorbing efficiency depends on the $\vec{k}$-vector of the incoming light. Although the claim that UPMLs do not reflect light is absolutely correct, when light travels through such media the absorption efficiency impairs as the angle of incidence grows, defined it with respect to the normal direction to the interface between the UMPL layer and the empty space. This becomes a source of inaccuracies for isolated systems if they are very elongated. In periodic systems it is also crucial to avoid reflection, specially that light scattered at almost grazing angles. This is important since in periodic systems a lot of interesting phenomena occur at wavelengths close to grazing angle condition. Let us illustrate this with an example. Figure 1.10 shows an schematic 1D-grating deposited on gold. In a periodic system, the diffracted waves are characterized by wavevectors satisfying

$$\vec{k}_{\parallel}^{n} = \vec{k}_{\parallel}^{i} + \frac{2\pi}{\Delta_x} n\hat{x} \tag{1.70}$$

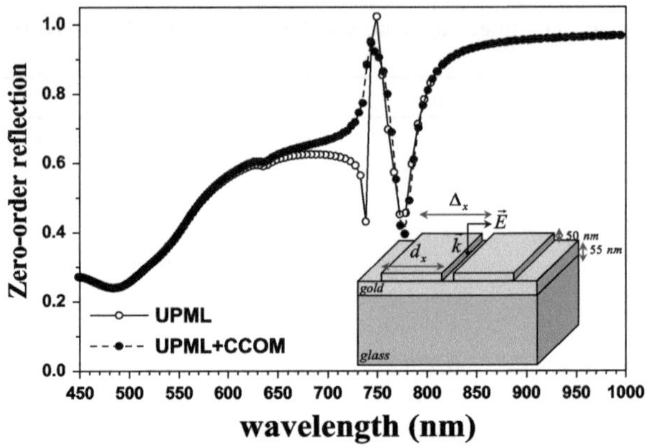

Fig. 1.10 Zero-order reflection spectra of a 1D-grating array (grating thickness = 50 nm) on a gold film (55 nm) computed using two different absorbing boundaries. The rest of geometrical parameters are: $\Delta_x = 740$ nm, $d_x = 370$ nm. Empty symbols render the calculation in which the CCOM+UPML absorbing boundaries are used jointly. The reflection peak at $\lambda \sim 725$ nm is due to spurious reflection at the "thin" UPML layer

where Δ_x represents the period of the array and, $\vec{k}^i_{\|}$ is the incident k-vector. Each diffracted mode has $k_z = \pm\sqrt{\varepsilon g^2 - \vec{k}^2_{\|}}$, $(g = \omega/c)$. Note that if $k_z \to 0$, $\vec{k}$ becomes grazing. For normal incidence this occurs at $\lambda_n = \frac{\sqrt{\varepsilon}\Delta_x}{n}$, and it is precisely at that condition one could expect that UPMLs do not properly work. In Fig. 1.10, empty symbols show the zero-order reflection spectrum when only UMPLs are used. At 740 nm wavelength there is a sudden jump in reflection. This jump is not physical. Interestingly this wavelength holds the condition $\lambda_n = \frac{\Delta_x}{n}$ for $n = 1$ (in the reflection half-space $\varepsilon = 1$). To overcome this, we could try to enlarge the UPMLs thickness until no reflection from them was observed. However, there is a better way to proceed. Besides the UMPLs boundary conditions, we have implemented a different sort of absorbing boundaries that complement the former. First introduced by Ramahi [23, 24], the basic premise of the so-called complementary concurrent operators method (CCOM), is that by applying a set of differential operators at the boundaries, the method is able to cancel outer-boundary reflections. The cancelation is possible by averaging two independent solutions to the modeling problem. These two solutions are obtained by imposing radiation boundary operators that are complementary to each other. It is out of the scope of this section to deal with the technical details (See [13] for further explanations). In any case, we would like to show the reader how by using the CCOM method is obtained the correct optical spectrum, which can be seen in Fig. 1.10 depicted with full symbols.

1.3 The Coupled Mode Method: An Overview

Modal expansion methods rely on representing the field in terms of a base of eigenfunctions for each different "region", in which the whole system is divided. The solution is obtained after applying the proper boundary conditions. These methods lay on the "a priori" knowledge of exact solutions in the different regions, considering that each one fills in the whole space. In this thesis, we will use a modal expansion method applied to electromagnetism problems. Namely, the CMM will be used in Chap. 2, Sect. 2.4 when investigating the Extraordinary Optical Transmission phenomenon, also in Chap. 3 where Negative Refractive Index presented on the double-fishnet structure is studied, and finally in Chap. 4, Sect. 4.2 for designing a source for SPPs. It is not pretended here to fully explain this technique for the aforementioned problems. We limit to briefly describe the general ideas behind the method instead.

The details can be found in several references. Let us mention some of them, in order to give an idea of the type of problems that can be treated with this method:

i. The phenomenon of EOT of light has been extensively investigated with CMM : through slit arrays [25–27], hole arrays [28–32] and also through quasi-periodic structures [33, 34].
ii. Optical response of single apertures are treated with the CMM approximation too. For example, it has been investigated resonant transmission through single holes [35, 36] and through finite chains of subwavelength holes [37]. The CMM approximation can treat, in some cases, nonlinear optical response as it was demonstrated in Ref. [38].
iii. Scattering of modes bounded to a surface (like SPPs) is another problem where CMM is able to reach good results. It has been used in investigating, for instance, the scattering of SPPs by one-dimensional periodic nano-structured surfaces [39] and in a fully 3D-problem, the scattering of light and SPPs by arrays of holes [40].
iv. The CMM approximation can also be applied to problems not related to optics, for instance, EOT mediated by surface sonic waves [41], and even the transmission of cold atoms through optically induced potential barriers [42].

To illustrate CMM, let us describe the theoretical formalism used for calculating optical spectra through 2DHAs drilled on PEC metallic films placed in symmetric environments, i.e., between a substrate and a cover optically characterized by the same dielectric constant, ε. An extended discussion of this problem can be found in Ref. [31], and references therein.

The whole space is divided in three regions (See Fig. 1.11). The EM fields, or rather, their components lying on the xy-plane are expanded in plane waves (region *I* and *III*), whereas in region *II* the fields are chosen to be the modes of an infinite "hole" shaped waveguide. Next, the fields have to be matched by means of the proper boundary conditions. More precisely, the EM-fields are expressed in terms of plane waves $|\mathbf{k}\sigma\rangle$, characterized by the in-plane component of the wave-vector

$\mathbf{k} = (k_x, k_y)$ and the polarization $\sigma = p$ or s. The representation of the modes in the dielectric half spaces then reads,

$$\begin{aligned}\langle \mathbf{r}|\mathbf{k}\, p\rangle &= (k_x, k_y)^T \exp(\imath \mathbf{k}\, r)/(Nk)\\ \langle \mathbf{r}|\mathbf{k}\, s\rangle &= (-k_y, k_x)^T \exp(\imath \mathbf{k}\, \mathbf{r})/(Nk)\end{aligned} \tag{1.71}$$

where $\mathbf{r} = (r_x, r_y)$, $k_z = \sqrt{\varepsilon g^2 - k^2}$ with $g = 2\pi/\lambda$ (T standing for transposition). The normalization constant $N = d$ for 2DHAs. Note also that H-fields are obtained from E-fields with the help of Maxwell's equations in the CGS system of units (so in a uniform media with dielectric constant ε, $\nabla \wedge \mathbf{E} = \imath g\mathbf{H}$ and $\nabla \wedge \mathbf{H} = -\imath \varepsilon g\mathbf{E}$), so that impedances (quotient between magnetic and electric fields) are dimensionless. For an incident plane wave with parallel wave vector $\mathbf{k}$ the in-plane EM fields in the reflection free space region (*I*) can be written as,

$$\begin{aligned}|\mathbf{E}_{\mathrm{I}}(z)\rangle &= |\mathbf{k}_0\sigma_0\rangle e^{ik_0 z} + \sum_{\mathbf{k}\sigma} r_{\mathbf{k}\sigma}|\mathbf{k}\sigma\rangle e^{-ik_z z}\\ |-\mathbf{u}_z \wedge \mathbf{H}_{\mathrm{I}}(z)\rangle &= Y_{\mathbf{k}_0\sigma_0}|\mathbf{k}_0\sigma_0\rangle e^{ik_0 z} - \sum_{\mathbf{k}\sigma} Y_{\mathbf{k}\sigma} r_{\mathbf{k}\sigma}|\mathbf{k}\sigma\rangle e^{-ik_z z}\end{aligned} \tag{1.72}$$

In this way, the EM fields in the transmission region (*III*) are

$$\begin{aligned}|\mathbf{E}_{\mathrm{III}}(z)\rangle &= \sum_{\mathbf{k}\sigma} t_{\mathbf{k}\sigma}|\mathbf{k}\sigma\rangle e^{ik_z(z-h)}\\ |-\mathbf{u}_z \wedge \mathbf{H}_{\mathrm{III}}(z)\rangle &= \sum_{\mathbf{k}\sigma} Y_{\mathbf{k}\sigma} t_{\mathbf{k}\sigma}|\mathbf{k}\sigma\rangle e^{ik_z(z-h)}\end{aligned} \tag{1.73}$$

Here $r_{\mathbf{k}}$ and $t_{\mathbf{k}}$ are the reflection and transmission coefficients and $\mathbf{u_z}$ is the unitary vector along the z-direction. On the other hand $Y_{\mathbf{k}s} = k_z/g$ and $Y_{\mathbf{k}p} = \varepsilon g/k_z$. Note that the $\mathbf{k}$-vector runs over the reciprocal lattice vectors. Usually a few of them are enough to obtain accurate results. Inside the holes, EM-fields are expanded in terms of the *TE* and *TM* waveguide eigenmodes. However, for subwavelength rectangular shaped holes good convergency is attained only considering the less decaying *TE* mode (the fundamental waveguide mode). Moreover, within this "minimal model" results can be worked out analytically, which greatly helps the physical insight. From now on, this mode is labeled as 0-mode. Thus we can write,

$$\begin{aligned}|\mathbf{E}_{\mathrm{II}}(z)\rangle &= |0\rangle[A_0 e^{iq_z z} + B_0 e^{-i\, q_z z}]\\ |-\mathbf{u}_z \wedge \mathbf{H}_{\mathrm{II}}(z)\rangle &= Y_0|0\rangle[A_0 e^{iq_z z} - B_0 e^{-iq_z z}]\end{aligned} \tag{1.74}$$

where the quantity $Y_0 = q_z/g$ corresponds to the admittance of the 0-mode, being the propagation constant of the fundamental mode $q_z = \sqrt{\varepsilon_{\mathrm{hole}} g^2 - q_y^2}$, ($q_y = \pi/a_y$). In this case, the fundamental waveguide mode in real space is: $\langle \mathbf{r}|0\rangle = (1, 0)^T \sin[q_y(y + a_y/2)]/\sqrt{M}$, $M = a_x a_y/2$ being a normalization factor, whether $\mathbf{r} \in [-a_y/2, a_y/2] \times [-a_x/2, a_x/2]$ and zero otherwise.

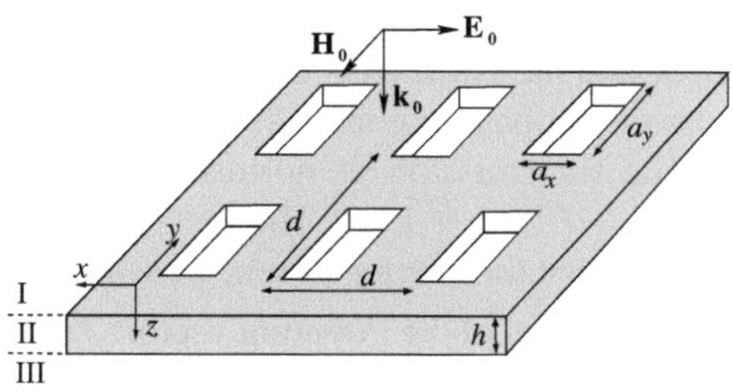

Fig. 1.11 Schematic picture of a square array of rectangular holes of side a_x and a_y perforated on a free-standing metallic film of thickness h. Parameter d defines the period of the array. The apertures are illuminated by a p-polarized plane wave at normal incidence

Note that when matching the EM fields at the interfaces ($z=0$ and $z=h$), E-field components parallel to the surface must be continuous over the whole surface. However H-field components parallel to the interfaces must be continuous *only* over the aperture, due to the presence of surface currents in the metal interface. This means that E-fields must be projected onto $\{|\mathbf{k}\sigma\rangle\}$ set of eigenvectors, whereas H-fields must be projected onto the fundamental waveguide mode [31]. Coupling between the 0-mode and free-space modes comes from overlap integrals, $\langle\mathbf{k}\sigma|0\rangle$, which depends upon the specific hole shape. For rectangular holes,

$$\langle\mathbf{k}\sigma|0\rangle = f_{\mathbf{k},\sigma}\sqrt{\frac{a_x a_y}{2d^2}}\,\mathrm{sinc}\,[k_x a_x/2]\left\{\mathrm{sinc}\left[\frac{(k_y+q_y)a_y}{2}\right]+\mathrm{sinc}\left[\frac{(k_y-q_y)a_y}{2}\right]\right\} \tag{1.75}$$

where $f_{\mathbf{k},p}=k_x/k,\ f_{\mathbf{k},s}=-k_y/k$ and $\mathrm{sinc}[x]=\sin[x]/x$.

By matching the EM fields at the interfaces, we end up with a set of two coupled linear equations for $\{E, E'\}$:

$$\begin{aligned}(G-\Sigma)E - G_V E' &= I_0\\ (G-\Sigma)E' - G_V E &= 0\end{aligned} \tag{1.76}$$

leading to the solution

$$E = \frac{I_0(G-\Sigma)}{(G-\Sigma)^2 - G_v^2},\quad E' = \frac{I_0 G_v}{(G-\Sigma)^2 - G_v^2} \tag{1.77}$$

where the expansion coefficients have been reorganized as follows, $E = A_0 + B_0$ and $E' = -(A_0 e^{iq_z h} + B_0 e^{-iq_z h})$. These new coefficients are the 0-modal amplitudes of the electric field at the input and the output sides of the holes, respectively.

The different terms of these equations have a simple physical interpretation. The term I_0 measures the overlap between the incident plane wave and the 0-mode inside the hole $I_0 = 2\imath Y_{\mathbf{k}_0,\sigma_0}|\langle\mathbf{k}_0\sigma_0|0\rangle|^2$. The term G_v in Eqs. 1.77 controls the EM coupling between the input and the output sides of the holes, $G_v = \frac{2\imath Y_0}{e^{2iq_z h}-1}$. The expression for the *self-energy* Σ is given by, $\Sigma = \imath Y_0 \frac{e^{2iq_z h}+1}{e^{2iq_z h}-1}$.

Finally, the EM-coupling between a hole and the continuum of plane waves, is mediated by the term G, which can be expressed as $G = \imath\sum_{\mathbf{k}\sigma} Y_{\mathbf{k}\sigma}|\langle\mathbf{k}\sigma|0\rangle|^2$. The real part of the Green function (G_r) controls the matching between the 0-waveguide mode and the evanescent EM modes in vacuum, and so does the imaginary part (G_i) with the propagating modes.

The transmission at normal incidence T relates to $\{E, E'\}$ coefficients from $T = G_i|E'|^2/\sqrt{\varepsilon}$ [31]. The last expression can be easily found if one takes into account that the energy current crossing a unit cell at a given z, $J(z)$, can be computed by integration of the Poynting vector: $J(z) = \frac{1}{2} Re\left[\int d\mathbf{r}\ \mathbf{u}_z(\vec{E}(\mathbf{r}, z) \wedge \vec{H}^*(\mathbf{r}, z)\right] = \frac{1}{2} Re\left[\langle -\mathbf{u}_z \wedge \mathbf{H}(z)|\mathbf{E}(z)\rangle\right]$. Using this expression, the incident current yields $J_0 = Y_{\mathbf{k_0}\sigma_0}/2$, thereby transmission in region *III* is then computed as $T = J_{\text{III}}/J_0$, where J_{III} is the current evaluated at coordinate $z > h$. After some algebra we arrive to an analytical formula for transmission

$$T = \frac{|I_o|^2}{4\sqrt{\varepsilon}} \frac{G_i}{(G_i)^2 + \left(\frac{|G-\Sigma|^2 - |G_v|^2}{2|G_v|}\right)^2} \tag{1.78}$$

CMM can take into account also realistic values for the dielectric constant of metals. The last is done by approximating the "penetration" into the metal surface by means of the surface-impedance boundary-conditions (SIBCs), except for the vertical walls of the holes which are treated as perfect conductor surfaces. Following the same procedure as for PEC metals it can be demonstrated the transmission formula (Eq. 1.78) holds also for real metals albeit redefining some quantities [37]:

$$\begin{aligned} I_0 &= \frac{2\imath Y_{\mathbf{k_0},\sigma_0}}{1 + Z_s Y_{\mathbf{k_0},\sigma_0}} |\langle \mathbf{k}_0\sigma_0|0\rangle|^2 \\ G_v &= \frac{2\imath Y_0}{e^{\imath q_z h}(1 + Z_s Y_0)^2 - e^{-\imath q_z h}(1 - Z_s Y_0)^2} \\ \Sigma &= \imath Y_0 \frac{e^{\imath q_z h}(1 + Z_s Y_0) + e^{-\imath q_z h}(1 - Z_s Y_0)}{e^{\imath q_z h}(1 + Z_s Y_0)^2 - e^{-\imath q_z h}(1 - Z_s Y_0)^2} \\ G &= \imath \sum_{\mathbf{k}\sigma} \frac{Y_{\mathbf{k}\sigma}}{1 + Z_s Y_{\mathbf{k}\sigma}} |\langle \mathbf{k}\sigma|0\rangle|^2 \end{aligned} \tag{1.79}$$

where $Z_s = 1/\sqrt{\varepsilon_m}$ (being ε_m the dielectric constant of the metal), is termed Surface Impedance. The 0-modal amplitudes of the electric field at the input and the output sides of the holes in this case are $E = A_0(1 - Z_s Y_0) + B_0(1 + Z_s Y_0)$ and $E' = -[A_0(1 + Z_s Y_0)e^{\imath q_z h} + B_0(1 - Z_s Y_0)e^{-\imath q_z h}]$.

We would like to pay a bit more attention over how Maxwell's equations are approximated within SIBC. In the *cgs* system of units and considering harmonically oscillating fields ($\sim e^{-\imath\omega t}$) inside metal

$$\vec{E} = \frac{i}{g\varepsilon_m} \nabla \wedge \vec{H} = \frac{i}{g\varepsilon_m} \begin{pmatrix} \partial_y H_z - \partial_z H_y \\ \partial_z H_x - \partial_x H_z \\ \partial_x H_y - \partial_y H_x \end{pmatrix}$$

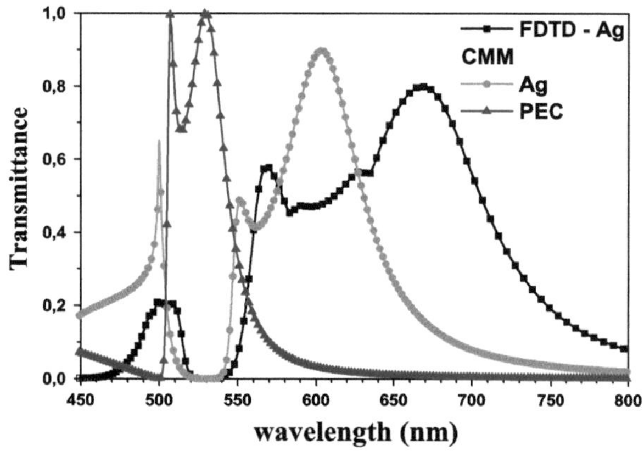

Fig. 1.12 Transmission spectra at normal incidence through an array of *square holes* ($d = 500$ nm, $a_x = a_y = 250$ nm and $h = 200$ nm) drilled on a free-standing silver film as calculated with FDTD and CMM. With CMM the two approximations described for simulating metal properties (PEC and SIBC) are shown

$$\vec{E}_{//} = \frac{i}{g\varepsilon_m}\begin{pmatrix}\partial_y H_z - \partial_z H_y \\ \partial_z H_x - \partial_x H_z\end{pmatrix} \simeq \frac{i}{g\varepsilon_m}\frac{\partial}{\partial z}\begin{pmatrix}-H_y \\ H_x\end{pmatrix} = \frac{i}{g\varepsilon_m}\frac{\partial}{\partial z}[\hat{u_z} \wedge \vec{H}_{//}] \tag{1.80}$$

owing to EM fields quickly decay into metals, the derivative of fields along the direction perpendicular to the surface is dominant over the other ones

$$\begin{aligned} |\partial_y H_z| &\ll |\, \partial_z H_y| \\ |\partial_x H_z| &\ll |\, \partial_z H_x| \end{aligned} \tag{1.81}$$

these are the so-called SIBCs that have been applied to obtain (1.80).

We are looking for EM solutions inside the metal in the form of

$$\begin{aligned} \vec{H}_{//}(\vec{r}) &= \vec{H}_o e^{i\vec{k}\vec{r}} \\ \vec{E}_{//}(\vec{r}) &= \vec{E}_o e^{i\vec{k}\vec{r}} \end{aligned} \tag{1.82}$$

where $\vec{k} = (k_{\|}, k_z)$. Taking into account (1.80) we obtain the approximated relationship between $\vec{E}_{//}$ and $\vec{H}_{//}$:

$$\begin{aligned} \vec{E}_{//}(\vec{r}) &\approx -\frac{k_z^m}{g\varepsilon_m}[\hat{z} \wedge \vec{H}_{//}(\vec{r})] \\ \vec{E}_o &\approx \frac{k_z^m}{g\varepsilon_m}[-\hat{z} \wedge \vec{H}_o] = Z_s(g)[-\hat{z} \wedge \vec{H}_o] \end{aligned} \tag{1.83}$$

Here $k_z^m = \sqrt{\varepsilon_m g^2 - k_{//}^2}$, and $\hat{z}$ is a unitary vector pointing perpendicularly to the surface, from outside to inside the metal.

Notice that, within SIBC the in-plane E-field is proportional to the in-plane H-field inside the metal. More importantly, since $\vec{E}_{//}$ is continuous at the interface,

$Z_s(g)[-\hat{z} \wedge \vec{H}_{//}]$ will be continuous too. This is the key to improve CMM from the PEC approximation to the SIBC approximation. In order to obtain Eq. 1.79 we must impose a "new" set of boundary conditions, at $z = 0$, namely $|\mathbf{E}_{\mathrm{I}}(z)\rangle - Z_s(g)| - \mathbf{u}_z \wedge \mathbf{H}_{\mathrm{I}}(z)\rangle$ must be continuous, while at $z = h$ the new boundary condition reads $|\mathbf{E}_{\mathrm{I}}(z)\rangle + Z_s(g)| - \mathbf{u}_z \wedge \mathbf{H}_{\mathrm{I}}(z)\rangle$. Note that $\hat{z} = -\mathbf{u}_z$ in the latter case.

Finally, we would like to discuss how accurately SIBC represents optical properties of metals. Usually, when dealing with the SIBC approximation k_z^m is approximated by $\sqrt{\varepsilon_m} g$, therefore $Z_s = 1/\sqrt{\varepsilon_m}$. However, in order to efficiently incorporate SPPs we could approximate $k_z^m = \sqrt{\varepsilon_m g^2 - k_{\mathrm{SPP}}^2}$ $(k_{\mathrm{SPP}} = g\sqrt{\frac{\varepsilon\varepsilon_m}{\varepsilon+\varepsilon_m}})$ obtaining an accurate surface impedance, $Z_s = 1/\sqrt{\varepsilon + \varepsilon_m}$.

In Fig. 1.12 we present an example of transmission spectra carried out with CMM within the "minimal model" approximation (only one waveguide mode inside holes) and we show how it compares with the "exact" FDTD calculation. The structure is an array of square holes defined by the parameters $d = 500$ nm, $a_x = a_y = 250$ nm and $h = 200$ nm. The agreement between FDTD and CMM is good. As expected, it is better for silver than for the PEC case. However, notice that these calculations are conducted under the following approximation: the hole walls are considered as PEC even when SIBC operates. In order to overcome the latter, in Fig. 1.12 the hole side is actually widened as much as one skin-depth to mimic the penetration of the EM fields into the metal walls. A better solution consists in using the propagation constant of the real metal waveguide, as we will see in Chap. 2.

References

1. V.M. Agranovich, D.L. Mills, *Surface Polaritons* (North Holland, Amsterdam, 1982)
2. A.D. Boardman, *Electromagnetic Surface Modes* (Wiley, New York, 1982)
3. H. Raether, *Surface Plasmons on Smooth and Rough Surfaces and on Gratings* (Springer, Berlin, 1988)
4. A.V. Zayats, I.I. Smolyaninov, A.A. Maradudin, Phys. Rep. **408**, 131 (2004)
5. S.A. Maier, *Plasmonics: Fundamentals and Applications* (Springer, Berlin, 2007)
6. S.I. Bozhevolnyi, *Plasmonic Nanoguides and Circuits* (Pan Stanford Publishing, Singapore, 2009)
7. T.W. Ebbesen, H.L. Lezec, H.F. Ghaemi, T. Thio, P.A. Wolff, Nature **391**, 667 (1998)
8. J. Zenneck, Ann. Phys. **23**, 846 (1907)
9. J.D. Jackson, *Classical Electrodynamics*, 2nd edn. (Wiley, New York, 1975)
10. E.D.Palik, in *Handbook of optical constants of solids* ed. by D.E. Palik. Academic Press Handbook Series (Academic Press, New York, 1985)
11. W.L. Barnes, J. Opt. A: Pure Appl. Opt. **8**, S87 (2006)
12. K. Yee, IEEE Trans. Antennas Propag. **14**, 302 (1966)
13. A. Taflove, S.C. Hagness, *Computational Electrodynamics: The Finite-Difference Time-Domain Method*, 3rd edn. (Artech House, Boston, 2005)
14. A. Belkhir, F.I. Baida, Phys. Rev. E **77**, 056701 (2008)
15. F.I. Baida, A. Belkhir, Opt. Lett. **34**, 2453 (2009)
16. C.T. Chan, Q.L. Yu, K.M. Ho, Phys. Rev. B **51**, 16635 (1995)
17. A.J. Ward, J.B. Pendry, Comput. Phys. Commun. **128**, 590 (2000)

18. J.D. Joannopoulos, S.G. Johnson, R.D. Meade, J.N. Winn, *Photonic Crystals: Molding the Flow of Light* (Princeton University press, Princeton, 2008)
19. R.J. Luebbers, F. Hunsberger, K.S. Kunz, R. Standler, IEEE Trans. Electromagn. Compat. **32**, 222 (1990)
20. R.J. Luebbers, F. Hunsberger, K.S. Kunz, IEEE Trans. Antennas Propag. **39**, 29 (1991)
21. R.J. Luebbers, F. Hunsberger, IEEE Trans. Antennas Propag. **40**, 1297 (1992)
22. J.P. Berenguer, J. Comput. Phys. **114**, 185 (1994)
23. O.M. Ramahi, IEEE Antennas Propag. Mag. **39**, 33 (1997)
24. O.M. Ramahi, IEEE Trans. Antennas Propag. **46**, 1475 (1998)
25. J.A. Porto, F.J. García-Vidal, J.B. Pendry, Phys. Rev. Lett. **83**, 2845 (1999)
26. J. Bravo-Abad, L. Martín-Moreno, F.J. García-Vidal, Phys. Rev. E **69**, 026601 (2004)
27. A.I. Fernández-Domínguez, F.J. García-Vidal, L. Martín-Moreno, Phys. Rev. B **76**, 235430 (2007)
28. L. Martín-Moreno, F.J. García-Vidal, Opt. Express **12**, 3619 (2004)
29. F.J. García-Vidal, L. Martín-Moreno, J.B. Pendry, J. Opt. A: Pure Appl. Opt. **7**, S97 (2005)
30. J. Bravo-Abad, A. Degiron, F. Przybilla, C. Genet, F.J. Garcia-Vidal, L. Martin-Moreno, T.W. Ebbesen, Nat. Phys. **2**, 120 (2006)
31. L. Martín-Moreno, F.J. García-Vidal, J. Phys.: Condens. Matter **20**, 3042141 (2008)
32. F.J. García-Vidal, L. Martín-Moreno, T.W. Ebbesen, L. Kuipers, Rev. Mod. Phys. **82**, 729 (2010)
33. J. Bravo-Abad, A.I. Fernández-Domínguez, F. J. García-Vidal, L. Martín-Moreno, Phys. Rev. Lett. **99**, 203905 (2007)
34. A.I. Fernández-Domínguez, I. Hernández-Carrasco, L. Martín-Moreno, F.J. García-Vidal, Electromagnetics **28**, 186 (2008)
35. F.J. García-Vidal, E. Moreno, J.A. Porto, L. Martín-Moreno, Phys. Rev. Lett. **95**, 103901 (2005)
36. A.Y. Nikitin, D. Zueco, F.J. García-Vidal, L. Martín-Moreno, Phys. Rev. B **78**, 165429 (2008)
37. J. Bravo-Abad, F.J. García-Vidal, L. Martín-Moreno, Phys. Rev. Lett. **93**, 227401 (2004)
38. J.A. Porto, L. Martín-Moreno, F.J. García-Vidal, Phys. Rev. B **70**, 081402 (2004)
39. F. López-Tejeira, F.J. García-Vidal, L. Martín-Moreno, Phys. Rev. B (R) **72**, 161405 (2005)
40. de F. León-Pérez, G. Brucoli, F.J. García-Vidal, L. Martín-Moreno, New J. Phys. **10**, 105017 (2008)
41. J. Christensen, A.I. Fernández-Domínguez, de F. León-Pérez, L. Martín-Moreno, F.J. García-Vidal, Nat. Phys. **3**, 851 (2007)
42. E. Moreno, A.I. Fernández-Domínguez, J.I. Cirac, F.J. García-Vidal, L. Martín-Moreno, Phys. Rev. Lett. **95**, 170406 (2005)

Chapter 2
Extraordinary Optical Transmission

2.1 Introduction

Anyone of us has experimented that light spreads in all directions upon interacting with objects. Another matter is how it does it. Diffraction theory is an old problem in optics, which goes back the works by Thomas Young and Augustin-Jean Fresnel in the nineteenth century. In 1944 an important landmark in that widely studied topic was put on the map by Hans Bethe [1]. He found that, at first approximation, the normalized-to-area transmission through a circular hole perforating an infinitely thin perfect conductor plate is

$$T \approx \frac{64}{27\pi^2}\left(\frac{r}{\lambda}\right)^4 \tag{2.1}$$

where λ is the wavelength of the incoming light, being r the radius of the hole. It was a great surprise indeed, because most of the well established theories at that time gave a $(r/\lambda)^2$ dependence. Furthermore he found the pre-factor, not only that unexpected dependence. Bethe's theory shows that a subwavelength hole ($\lambda \gg r$) is a poor device for transmitting light through.

Therefore, it is not strange that the discovery of the phenomenon of extraordinary optical transmission (EOT) through subwavelength holes by Ebbesen et al. [2] has been one of most important findings in the field of Optics in the last years. The basic structure in which EOT phenomenon emerges is a two-dimensional periodic array of subwavelength holes (2DHA) perforated on an optically thick metallic film. This phenomenon is characterized by the appearance of a series of transmission peaks and dips in the transmission spectrum. It is commonly accepted EOT occurs when the normalized to area transmission is larger than unity. On the other hand, when the transmission per hole in an array is larger than for an isolated hole is called enhanced transmission. Here we will not differentiate between these two cases. From the beginning, it was realized that the spectral locations of those resonant features coincide with the corresponding ones of surface plasmon polaritons (SPPs) [3]. This link between EOT and SPPs has been corroborated by subsequent theoretical works

S. G. Rodrigo, *Optical Properties of Nanostructured Metallic Systems*,
Springer Theses, DOI: 10.1007/978-3-642-23085-1_2,

[4, 5] and now it is widely accepted that the excitation of those surface electromagnetic (EM) modes is at the origin of EOT.

The EOT mechanism has sparked considerable interest for its fundamental implications and also from the applied point of view, as many potential applications based on this phenomenon have been proposed [6]. EOT observed in symmetrically perforated thin metal films [2, 4, 7, 8], the squeezing of the optical near-field by plasmon coupling resulting in focusing light into very small volumes [9], and beaming of light via a single slit (or hole) in thin metal films surrounded by a grating like structure [10] are only a few of many interesting examples. A vast number of applications have been suggested and some are currently in use, e.g., wavelength tunable filters, subwavelength lithography, near-field microscopy, surface enhanced Raman spectroscopy, etc. Obviously, it is out of the scope of this chapter to review so wide field of research. We recommend for further reading Ref. [6, 11, 12], and references therein.

Throughout this chapter, we just summarize a part of the contributions to EOT done during the course of this thesis [13–16]. In these works our aim was trying to understand some important aspects of the EOT when the parameters defining the structure presenting EOT are varied. It what follows we will discuss how EOT depends on: the metal chosen, the shape of the holes and the film thickness. The study will focus on the optical response of 2DHAs drilled on metal films. Additionally, we will investigate a quite different system that also displays EOT. The chapter is organized as follows:

a. In Sect. 2.2, we present a theoretical study, based on the finite difference time domain (FDTD) method, of the optical response of circular hole arrays drilled in several metal films (Ag, Au, Cu, Al, Ni, Cr and W). Two series of structures are studied. In the first one, transmittance peaks are analyzed as all geometrical parameters defining the system are scaled, except the metal thickness which is kept constant, showing a good agreement with existing experimental data. In the second series, also the metal thickness is scaled. These is no available experimental data for this case, but its theoretical consideration allows a clear distinction in the behavior of different metals.
b. A theoretical study is developed on the optical transmission through square hole arrays drilled in optically thin films in Sect. 2.3, by means of the FDTD method. Nano-structures containing thin films are interesting because transmission may occur through both the holes and the metal layer. Moreover, the EM bounded modes supported by thin films are not the same that those supported by thick films.
c. It is known two mechanisms leading to enhanced transmission of light in 2DHAs: excitation of SPPs and localized resonances that are also present in single holes. In Sect. 2.4 we analyze theoretically how these two mechanisms evolve when the period of the array is varied. Along with the FDTD method this work was also done with the aim of the coupled mode method (CMM). This method was adapted to this problem by Dr. A. Mary at the Departamento de Física Teórica de la Materia Condensada—Universidad Autónoma de Madrid.

d. Finally, in Sect. 2.5, the spectral dependence of the extraordinary transmission through monolayers of close-packed silica or polystyrene microspheres on a quartz support, covered with different thin metal films (Ag, Au and Ni) is investigated. Measured spectra are compared with modeled transmission spectra using FDTD calculations. The optical response of this system shows remarkable differences as compared with the "classical" 2DHA configuration.

2.2 Influence of Material Properties on EOT Through Hole Arrays

Pioneering attempts to understanding EOT pointed out to SPP modes [2, 4, 17, 18] as responsible for the phenomenon. More generally, it has been shown that EOT-like behavior occurs whenever two surface modes are coupled between themselves and weakly coupled to a continuum [4], allowing a Fano-like description of the process [19]. Examples of this general mechanism are the cases of wave transmission aided by: Brewster-Zenneck waves in hole arrays drilled in Tungsten [20], guide cavity modes on slit arrays covered by a dielectric layer [21], surface electromagnetic waves in photonic crystals [22–24], surface sonic waves [25], and even the transmission of cold atoms through optically induced potential barriers [26].

Back to the case of 2DHAs, many studies have been devoted to study the dependence of EOT on the different parameters defining the system. As regards to geometrical parameters, it has been found that the hole shape can strongly influence both the polarization properties and the intensity of the transmission. This has been related to the presence of single-hole transmission resonances that couple to the SPPs [15, 27–31].

With respect to the material properties, it is known that some metals (notably Au and Ag) are best suited for EOT than others (Ni, Co,...) [2, 3]. However, a systematic comparative between different metals was lacking until the experiments performed by Przybilla et al. were reported [32]. These experiments analyzed EOT through 2DHA made of circular holes drilled in optically thick metal films, deposited on a glass substrate. The study considered different metals and analyzed the peak transmittance as a function of lattice parameter which, in turn, controlled the resonant wavelength. If the system were a perfect electric conductor (PEC) the transmission would not depend on lattice parameter, provided all length scales in the system are scaled in the same way. Therefore, deviations from this behavior reflects the effect of material properties. In the experiments [32] the hole radii were scaled with the lattice parameter but, due to practical limitations, the metal thickness was kept constant, which makes the analysis even more complex. Another possible complication is that finite size effects, surface quality and imperfections in hole shape in a real system could depend on the metal considered.

In this section we present a theoretical study of EOT in periodic hole arrays drilled in different metals. First we compare with the experimental results presented in Ref. [32]. Additionally, we present a study in which *all* lengths are scaled with the lattice parameter. In both cases, comparison with the PEC case helps to understand the effect of material properties in the light transmission through 2DHA.

2.2.1 Theoretical Approach

Our calculations are performed with the FDTD method (See Sect. 1.2). Infinite periodic 2DHA are simulated by applying Bloch conditions at the boundaries of the unit cell and imposing "uniaxial perfect matched layers" at surfaces parallel to the metal film. In the FDTD method Maxwell's equations are discretized in both space and time. Therefore, convergence depends on both mesh size and temporal step. In order to properly calculate the influence of material properties, the rapidly decaying fields inside the metal should be accurately computed. This, together with the proper representation of circular holes in cartesian coordinates (which are the natural choice for square arrays in a film), impose very small mesh sizes. In this section we use mesh sizes ranging from 2 to 5 nm. The dielectric constant in cells at the metal-dielectric interface is taken as that of the medium with largest volume inside that particular cell.

For the calculation of the transmittance, the structures were excited by a gaussian wave-packet composed of normally incident plane-waves (with the electric field pointing along one of the axes of the square array) and all frequencies of interest (Sect. 1.2.2). Spectra were calculated after projection onto diffracted modes (Sect. 1.2.3). In the comparison with experimental data, only the zero order mode was considered in the post-processing, as experimental intensities were collected in a small solid angle centered around the normal direction. Additionally, in our study we will correlate transmittance features to the dispersion relation of modes supported by the corrugated slab. For this, the band structure is calculated exciting the system with a superposition of Bloch's waves with a well defined crystal momentum, and imposing Bloch's theorem at the boundaries (Sect. 1.2.3).

Metals require a special treatment in FDTD method because of the dielectric constant is local in frequency domain but non-local in the time-domain, as explained in Sect. 1.2.4. We have used the "piece linear recursive convolution method", which can efficiently treat dispersive media, provided their dielectric constant $\varepsilon(\omega)$ can be expressed as a sum of Drude and Lorentz terms:

$$\epsilon(\omega) = \epsilon_r - \sum_j \frac{\omega_{Pj}^2}{\omega(\omega + \imath\gamma_j)} - \sum_j \frac{\Delta\epsilon_j \Omega_j^2}{\omega^2 - \Omega_j^2 + \imath\omega\Gamma_j}. \tag{2.2}$$

We have considered the following materials: Ag, Au, Cu, Al, Ni, Cr, and W, taking the values for the parameters in the Drude-Lorentz form either from the literature (when available) [33, 34] or from fits to data in Palik's handbook [35, 36]. The fitting parameters used were given in Table 1.1, and the wavelength dependence of the dielectric constant obtained from them is depicted in Fig. 2.1 for reference. Also for

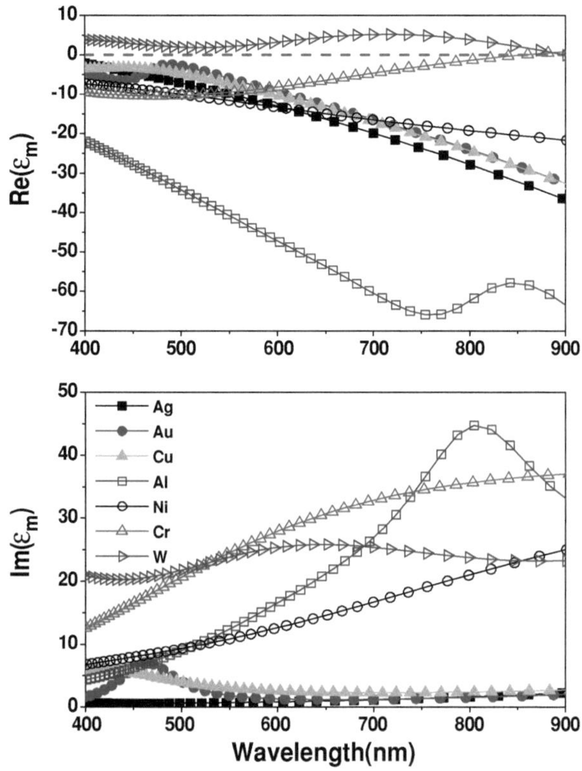

Fig. 2.1 Dielectric constant for Ag, Au, Cu, Al, Ni, Cr and W, as a function of wavelength (obtained with Eq. 2.2 from data in Table 1.1)

future reference, Fig. 2.2 renders both the skin depth, $\delta = [Im(k_m)]^{-1}$ (with $k_m^2 = \varepsilon_m \omega^2/c^2$) and the SPP absorption length, $L_{SPP} = [2Im(k_{SPP})]^{-1}$, being $k_{SPP} = (\omega/c)(\varepsilon_S \varepsilon_m/(\varepsilon_S + \varepsilon_m))^{1/2}$ the SPP longitudinal wavevector. Here ε_m and ε_S are the metal dielectric constant and the substrate dielectric constant, respectively. We stress here that the dielectric constants used in this study are taken directly from experimental data on bulk. No attempt has been made to improve the comparison between computed and experimental transmission spectra by incorporating additional fitting parameters.

2.2.2 *EOT Peak Related to the Metal-Substrate Surface Plasmon*

Peak Position

Figure 2.3 renders the FDTD results for the transmittance spectra of an array (lattice parameter $P = 400$ nm) of circular holes with diameter $d = P/1.75 \sim 230$ nm in different metal films. In all cases, the metal thickness is $w = 250$ nm, and the films

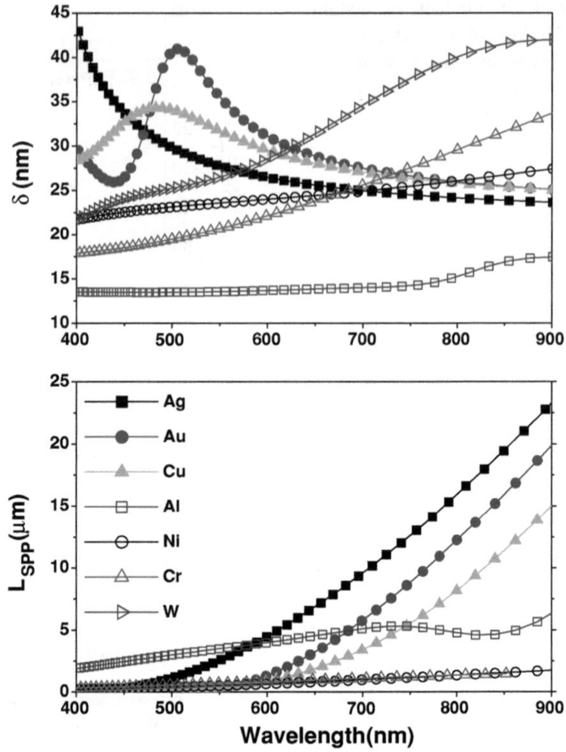

Fig. 2.2 Spectral dependence for different metals of (**a**) skin depth for a plane wave impinging at normal incidence on the metal surface (**b**) absorption length for SPP on the metal-dielectric interface

are deposited on a substrate with dielectric constant $\varepsilon_S = 2.25$. The figure clearly shows the difference in transmittance spectra between different metals. We also include the calculation for W which, in the considered frequency range, is a dielectric, i.e., $Re(\varepsilon_W) > 0$. Notice that the heights of EOT peaks in W are even smaller than those of the "bad" metals (Ni and Cr). Notice also that W has a transmission maximum very close to the Rayleigh condition, $\lambda_R = \sqrt{\varepsilon_S}P$, wavelength at which the $(1, 0)$ diffraction order changes character from evanescent to radiative, while the corresponding one in the metallic case is red-shifted (by as much as $\approx$125 nm in the case of the "good" metals in the optical regime: Ag, Au, Cu). In the rest of this section, and following Ref. [32], EOT is characterized by the transmittance peak appearing close to λ_R. More precisely, in the metallic case, this peak is related to the spectral location of the SPP of the corrugated structure [4, 32, 37]. Here we will label this peak as $S_{1,0}$. Figure 2.4 shows the $S_{1,0}$ peak spectral position (defined as $\lambda_{1,0}$) as a function of the period, for the parameters considered in Ref. [32] ($w = 250$ nm; $d = P/1.75$). In addition, Fig. 2.4 also renders the light line in the substrate (continuous line) and the results obtained by considering the metal as a PEC, i.e. a metal with $\varepsilon = -\infty$, with the same nominal parameters (asterisk data points). Notice that a flat metal surface does not support EM modes, but a periodically corrugated one behaves as if it had a Drude-like dielectric response in which the effective plasma frequency depends only

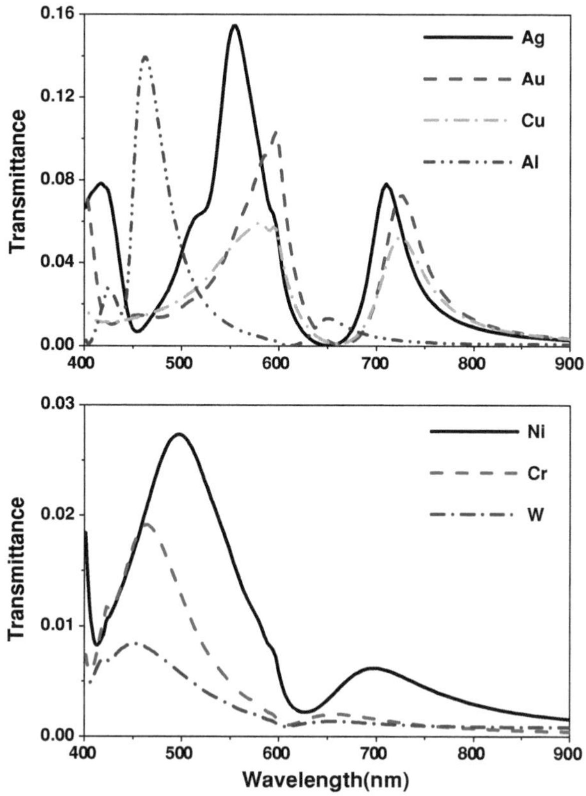

Fig. 2.3 A representative case of computed transmission spectra, for the different metals considered. The geometrical parameters defining the array are: period $P = 400$ nm, metal thickness $w = 250$ nm and hole diameter $d = P/1.75$

on the geometrical parameters [38–40]. Thus, comparison with the PEC case allows to discern geometrically induced effects from the ones due to material properties. In any case, the agreement between these calculations and the experimental measurements (see Fig. 2.3 in Ref. [32]) is remarkable. The small differences could be attributed to variations in the actual dielectric constant from the bulk value, to finite size effects [41], and/or to irregularities on hole shape and size related to the small uncertainties from focused ion beam (FIB) lithography technique (which has an accuracy of the order of 10 nm). In our opinion, the agreement obtained validates FDTD calculations as a predictive tool in this kind of systems. These results also confirm that, in the experiments considered, the effect of possible inhomogeneities in hole shape and size was not relevant. Notice that the good agreement with the experimental data was obtained by using the bulk dielectric constant, despite the fact the dielectric properties at the surface could have been modified by the processing related to drilling of the holes.

In order to obtain further insight on the origin of EOT phenomena, Fig. 2.5 renders the dependence with period of $\lambda_{1,0}$ for the metals Ag and Al, and also for W. In this figure, we have also included the spectral location of the minimum which appears

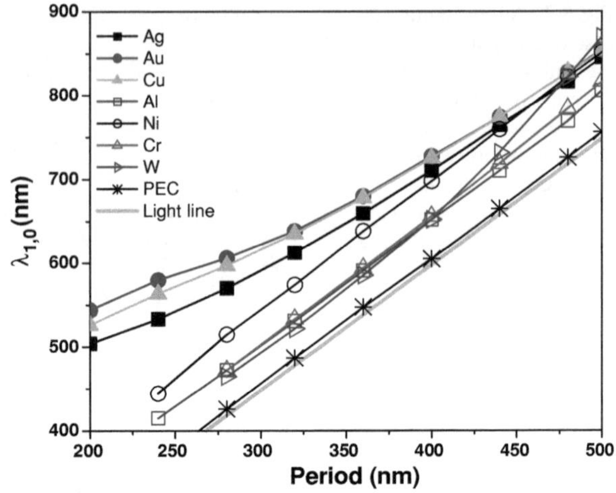

Fig. 2.4 (1,0) substrate peak position, $\lambda_{1,0}$, as a function of the array period, both for the metals investigated and for a perfect electrical conductor. Hole radius is scaled with the period as $d = P/1.75$ but film thickness is kept constant at $w = 250$ nm

associated to the $S_{1,0}$ peak, slightly blue-shifted from the maximum. Results are presented for the case of constant film thickness, $w = 250$ nm. In the same figure we represent both λ_R and the folded dispersion relation for the SPP of a *flat* metal-dielectric interface (given by the expression $(\omega/c)(\varepsilon_S\varepsilon_m/(\varepsilon_S + \varepsilon_m))^{1/2} = 2\pi/P$).

In the case of metals, the correspondence between transmittance minima and SPP of the flat surface (with no holes) is evident from the figure, with the maxima following the same trend at slightly longer wavelengths. In the case of the dielectric (W), both maximum and minimum have much smaller amplitudes with respect to the transmission background than in the case of metals (see Fig. 2.3), with the very weak minimum appearing at λ_R. Both maximum and minimum are related to the surface EM modes of the corrugated structure. More precisely, each surface EM mode has associated a maximum and minimum transmission, characterized by a resonant Fano-like function times a smooth function related to the coupling of the incident wave with the surface mode. At normal incidence, the incident wave can couple to SPPs in the periodically corrugated structure which originate from plane waves differing from $\vec{k}_{\parallel} = 0$ by a reciprocal lattice vector. This is confirmed by Fig. 2.6, which renders the folded light line, the folded SPP dispersion relation of the flat metal-vacuum interface, and the FDTD calculation for the band structure of surface EM modes of the corrugated surface.

Transmission Intensity

Up to here we have concentrated on the dependence of the spectral position of transmission anomalies with the period of the array. The results for the peak intensities are presented in Fig. 2.7, which renders the maximum transmittance of the $S_{1,0}$ peak, $T_{1,0}$, as a function of its spectral position (i.e. for different periods) for several materials. In panel (a) all lengths defining the system are scaled with the period,

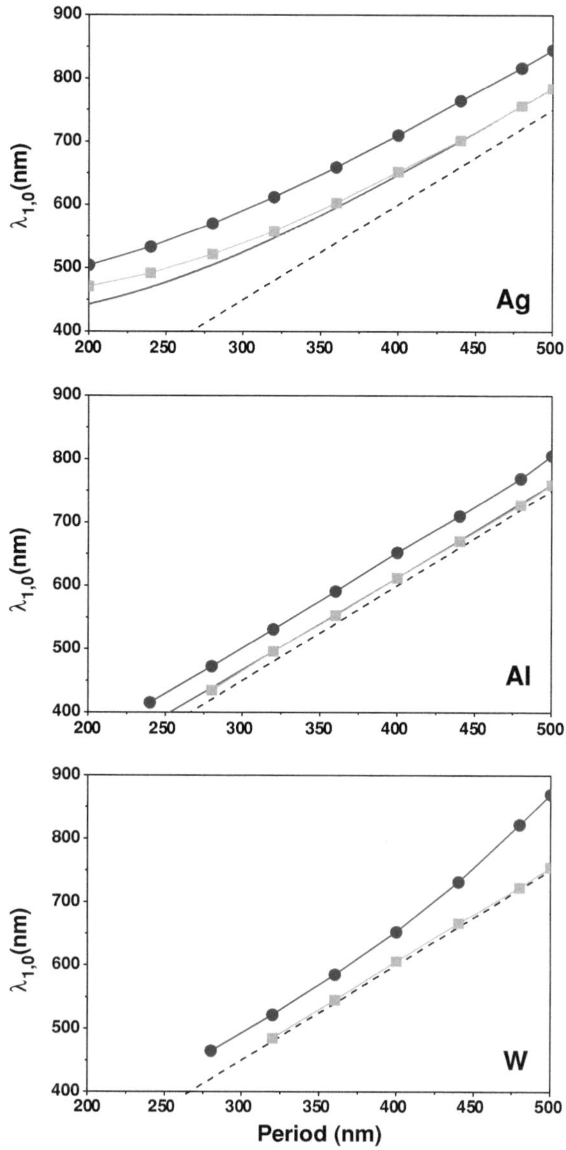

Fig. 2.5 Dependence on period of signatures related to the (1,0) substrate peak: maximum (*circular symbols*) and minimum (*square symbols*). Hole radius is scaled with the period as $d = P/1.75$ but film thickness is kept constant at $w = 250\,\text{nm}$. The *solid red line* represents the SPP dispersion relation for a flat metal-dielectric interface, while the *dashed black line* corresponds to the dielectric light line

except the metal thickness, which is kept constant at $w = 250$ nm (this is the case considered in Ref. [32]).

In panel (b) all lengths are scaled with the period. In this latter case, if the metal were a PEC, the peak transmittance would not depend on period (line with asterisks). A real metal presents two main differences with respect to a PEC, each of them having an opposite effect on the transmittance. On the one hand, real metals absorb energy,

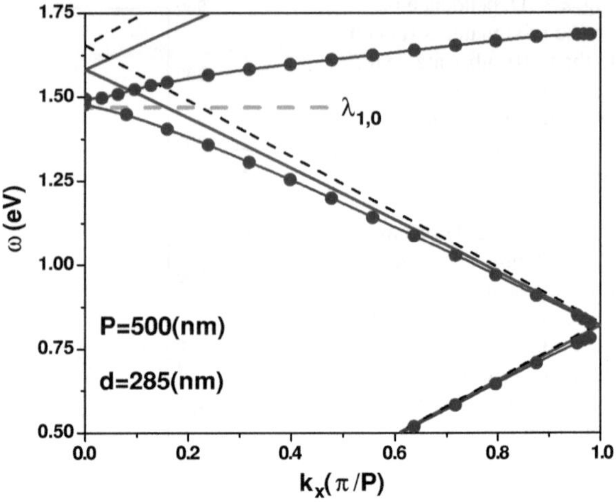

Fig. 2.6 SPP dispersion relation in the $\Gamma - X$ direction of the first Brillouin zone, for the dielectric-silver interface. *Circular symbols* (*blue line*) show the FDTD band structure for a 2DHA with period $P = 500$ nm and metal thickness $w = 250$ nm. Flat surface SPP dispersion relation is depicted with a *solid red line*. *Dashed black line* represents the light line. The spectral position of the (1,0) transmission maximum, $\lambda_{1,0}$, is depicted by a *dashed horizontal green line*

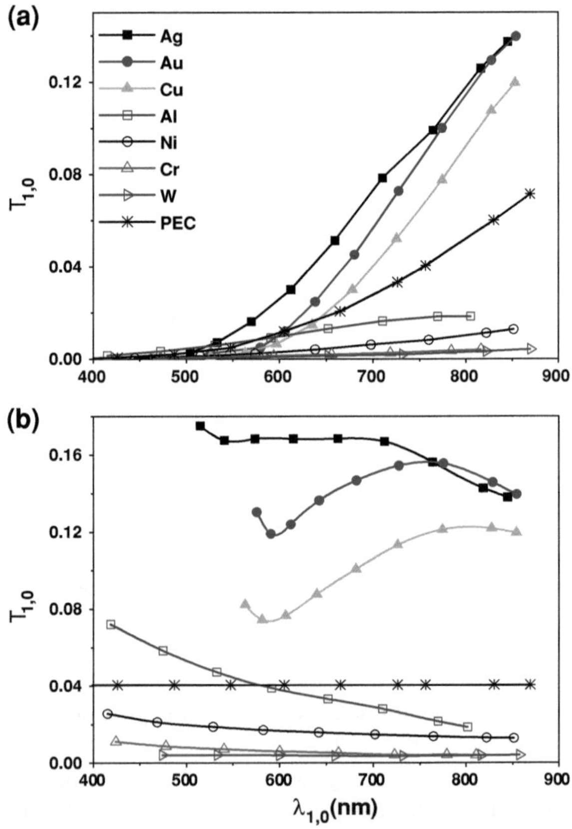

Fig. 2.7 Peak transmittance intensity, $T_{1,0}$, as a function of the peak spectral position, $\lambda_{1,0}$ for different metals and lattice periods. In both panels, hole diameter is scaled with period as $d = P/1.75$. In (**a**) metal thickness is kept constant at $w = 250$ nm. In (**b**) w is also scaled as $w = P/2.0$

which reduces the transmittance. This is more apparent in resonant processes, which require the EM field to stay for a longer time at the surface. As this resonance time is inversely proportional to the peak width, absorption has a larger influence on the narrowest transmittance peaks. On the other hand, the EM field penetrates in a real metal, effectively increasing the hole area accessible to the field and, therefore, increasing the transmittance. Several approximations can be envisaged in order to take into account the effective hole area. For instance, the hole can be considered as a finite portion of a waveguide. Effective areas can then be related to the propagation constants and the EM fields of different modes in the waveguide [42]. A simpler, phenomenological, approach is implemented by enlarging the hole radius by a factor (of order unity) times the skin depth [4]. This is a good approximation, provided the correction (skin depth) is much smaller than the hole radius. In any case, the effective hole area depends on the dielectric constant which, in turn, depends on wavelength. Therefore, from a electromagnetic point of view, even when all nominal lengths are scaled with the period, the "effective" surface percentage covered by holes does not remain constant. In what follows, we will show how the comparison with the PEC case allows to distinguish which of these two competing mechanisms (absorption and enlargement of the effective area) dominates for a particular circumstance. Let us start with the case of silver. The computed peak transmittance for silver is even larger than the corresponding one in a PEC *with the same nominal parameters*, as shown in Fig. 2.7. This suggest that, in this case, the "enlargement of effective area" mechanism is more important than absorption. In order to confirm this point, we have computed the transmittance peak intensities for 2DHA in PEC, but with the hole radius enlarged by the skin depth in Ag (computed at the wavelength at which the peak appears). We will refer to this as the "Corrected-PEC model". Figure 2.8 renders the results of these calculations, as well as the corresponding ones for 2DHA in Ag. Discontinuous lines represent the case in which all lengths in the system have been scaled with the period, except the metal thickness, which has been kept constant at $w = 250$ nm. The continuous lines render the case where also the metal thickness has been scaled with the period, as $w = P/2$. In all cases, the difference in transmission peak intensities between 2DHA in Ag and the enlarged holes in PEC is less than 30%. Notice that the Corrected-PEC model provide much more accurate results at large periods than at smaller ones, which can be associated to the increasing effect of absorption occurring at shorter periods and, correspondingly, smaller wavelengths. This is further corroborated by the fact that, at small periods, the Corrected-PEC model overestimates the peak transmittance.

This analysis helps understanding the results presented in Fig. 2.7b, and therefore the relative importance of the previously described mechanisms for different metals. At peak positions larger than $\lambda_{1,0} \approx 700$ nm, the dependence of $T_{1,0}$ on $\lambda_{1,0}$ for Au is similar to that of Ag, reflecting their similar skin-depths and absorption lengths. As $\lambda_{1,0}$ decreases, the difference in skin-depths in Au and Ag remains constant, but the ratio of their absorption lengths decreases. Correspondingly, due to absorption, $T_{1,0}$ in Au is smaller than that in Ag in this frequency regime. At the shorter $\lambda_{1,0}$ computed for Au this decreasing tendency is reversed, reflecting the increase in the skin-depth of Au (notice that the flattening out of the $\lambda_{1,0}$ versus period curve prevents exploring shorter

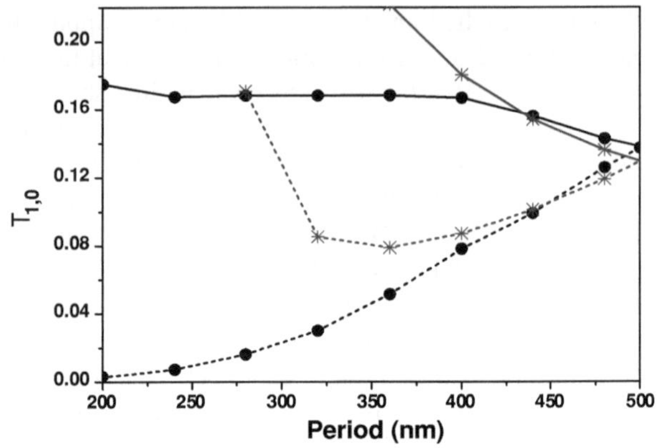

Fig. 2.8 Peak transmittance, $T_{1,0}$, as a function of the lattice period for silver (*circular symbols*) and PEC (*asterisks*). *Solid line* represents the case of scaled thickness ($w = P/2.0$), while the *dashed line* is for constant $w = 250$ nm. In the PEC case, the hole radius was enlarged by one silver skin depth (Fig. 2.2), evaluated at the corresponding wavelength

values of $\lambda_{1,0}$, see Fig. 2.4. To summarize, optical transmissions through 2DHA in Ag and Au are similar: in both cases the resonant transmission is larger than in a PEC with the same nominal parameters, the effect being caused by field penetration inside the metal, which effectively enlarges the hole area. Absorption is not the main factor for these metals. On the other hand, 2DHA in Cu have transmittance peak characteristics similar to those in Ag or Au, but with smaller values, reflecting the smaller absorption lengths in Cu than in the other two noble metals analyzed.

A completely different behavior occurs for a 2DHA in Ni or Cr: in both cases, the peak transmittance is always much smaller than that in Ag (or in a PEC). This behavior occurs although the skin depth in these metals can even be larger than in Ag, and is due to the large absorption present in both Ni and Cr.

Hole arrays in Al have transmittance peak characteristics more similar to the PEC case, reflecting the fact that the skin depth in Al is, at optical frequencies, much smaller than those of the other metals. Even so, $T_{1,0}$ depends on $\lambda_{1,0}$ for 2DHA in Al: the skin depth in Al remains approximately constant with wavelength, implying an "effective area" correction which is relatively smaller at larger hole areas (i.e., larger periods and larger $\lambda_{1,0}$ in the case in which all *nominal* lengths are scaled). This explains why, in this case, $T_{1,0}$ decreases with $\lambda_{1,0}$ even if the absorption length slightly increases (see Fig. 2.2) Notice that this behavior of the absorption length in Al is also very different to that in noble metals, where the absorption length increases strongly with wavelength. As a consequence, the absorption length in Al is, at $\lambda \approx 750$ nm, smaller than that in noble metals. This, combined with the unusually small skin depth makes that $T_{1,0}$ in Al has values of the order of those of "bad metals" (Cr, Ni) at the higher end of the spectral window considered. Finally, a 2DHA perforated in W present a very different transmission spectrum. As previously stated, in the considered spectral range W is a lossy dielectric. Transmission resonances are aided by Zenneck waves [20, 43], which are more weakly bound than SPPs. The resonances are much weaker, with a much smaller "visibility": the minima are not very deep and the maxima are not as high as in the case of metals (see Fig. 2.9). Also,

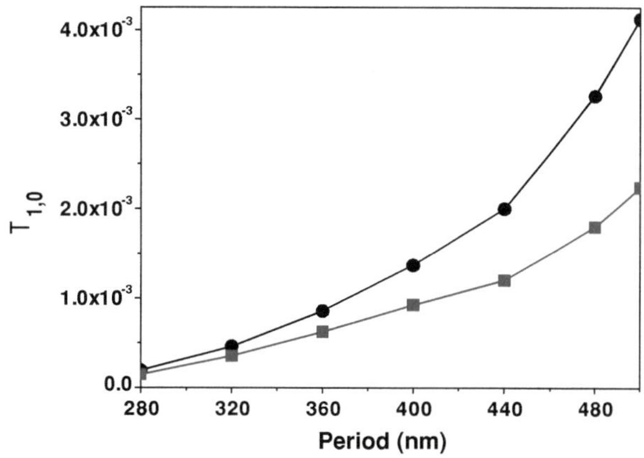

Fig. 2.9 Dependence on period of features of the $S_{1,0}$ peak in W: Maximum (*circular symbols*) and minimum transmittance (*square symbols*). All lengths have been scaled as in Fig. 2.7a

as mentioned before, the spectral position of $T_{1,0}$ in W (appearing very close to λ_R) is different to that of the metallic case.

Full-Width-at-Half-Maximum

The previous analysis on the relative importance of the skin depth and absorption length on the transmittance through 2DHA, based on the peak intensities, is reinforced by the spectral dependence of the peak linewidth. Figure 2.10 renders $\Gamma_{1,0}$, defined as the the full width at half maximum (FWHM) of the $S_{1,0}$ transmittance peak, as a function of $\lambda_{1,0}$. The calculations were performed for the configuration where all lengths are scaled, except the metal thickness which is kept fixed at $w = 250$ nm. Agreement between experiment [32] and theory is quite good, although experimental peaks are wider due to finite size effects and/or sample imperfections. The behavior of $\Gamma_{1,0}$ for a 2DHA in Au, Ag and Cu is similar. In this case, and for the parameters considered, radiation is the main loss channel for the surface EM mode, absorption being a (non-negligible) correction. Absorption is responsible for both the difference in $\Gamma_{1,0}$ between different metals and the decrease of $\Gamma_{1,0}$ with $\lambda_{1,0}$ (following the increase of the absorption length with wavelength). In the cases of 2DHA in either Cr or Ni, $\Gamma_{1,0}$ is much larger than that for noble metals, as expected, given that absorption lengths are much smaller in the former cases. Again, the case of Al is quite different from that of other metals: it goes from presenting the narrowest resonant peaks at small $\lambda_{1,0}$ to having values of $\Gamma_{1,0}$ of the order of those of "bad metals" for the larger $\lambda_{1,0}$ considered.

2.3 EOT Through Hole Arrays in Optically Thin Metal Films

As we have stated in the last section, since the discovery of EOT [2], numerous works have explored different parameter configurations of 2DHAs [6]. In the now

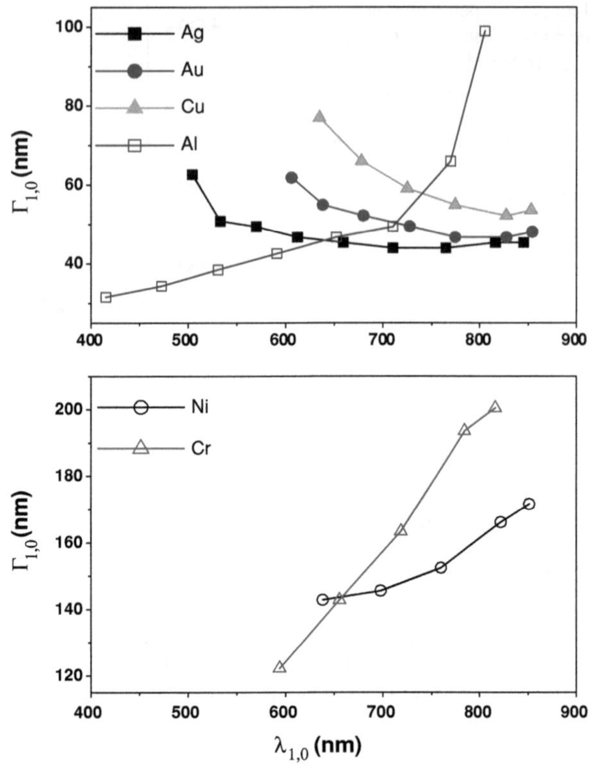

Fig. 2.10 Full width at half maximum for the $S_{1,0}$ peak, $\Gamma_{1,0}$, for 2DHAs in different metals. All lengths have been scaled as in Fig. 2.7a., i.e. the metal thickness is kept constant at $w = 250$ nm)

"canonical" configuration [2] the metal film is opaque. In this case, the EOT process involves surface modes at each side of the film which couple through the holes [4]. On the other hand, continuous metal films (thin enough to be translucent) also present transmission resonances when periodically corrugated. In this configuration, resonant spectral features are related to SPPs of the thin film [44, 45], the so called Short Range SPPs (SRs) and Long Range SPPs (LRs) [46].

The transmission of electromagnetic radiation through 2DHA, for thicknesses of the metal film ranging from less than 1 to 2–3 skin depths has been studied in the THz regime [47, 48]. These works showed how the intensity of the EOT peak developed with metal thickness, its spectral position being mainly determined by the lattice parameter. In this section, we extend the study to the optical regime. We analyze the optical response of 2DHAs on metal thickness, w, going from optically thick films to films as thin as approximately one "skin depth" (~ 20 nm).

To provide mechanical stability, actual thin films must lie on a substrate, which we take to be glass. We consider two different dielectric configurations: the asymmetric ($\varepsilon_I = \varepsilon_{II} = 1.0$; $\varepsilon_{III} = 2.25$) and the symmetric one ($\varepsilon_I = \varepsilon_{II} = \varepsilon_{III} = 2.25$), which can be experimentally obtained by using an index matching liquid. Throughout this section we consider square lattices of square holes; the period, P, is chosen to be

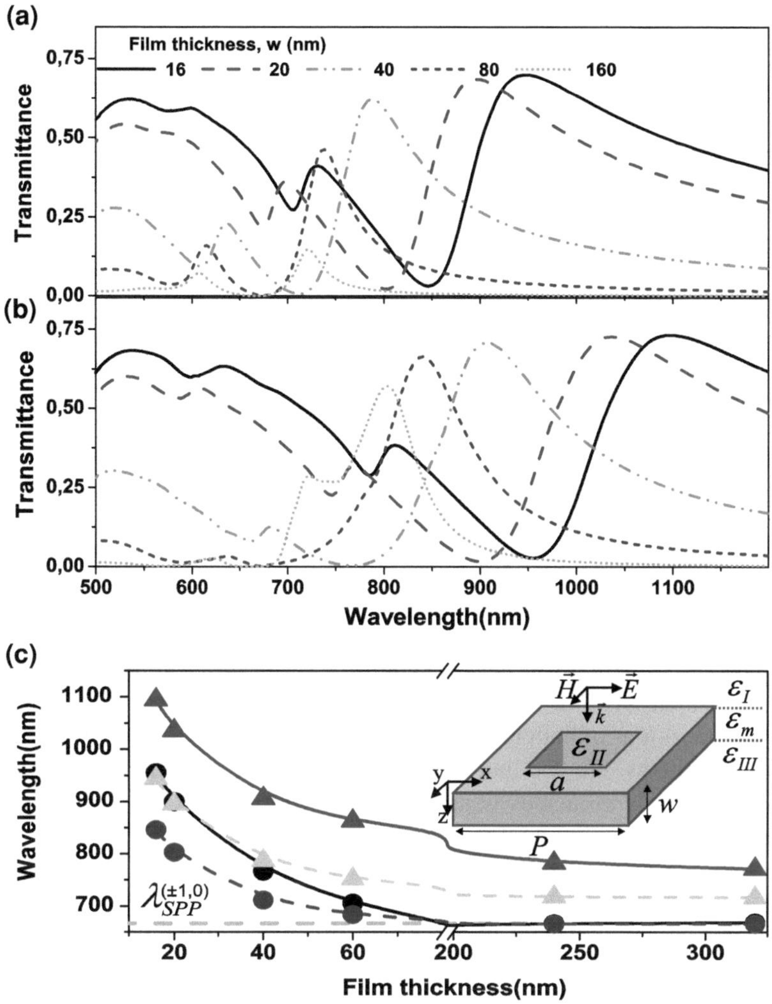

Fig. 2.11 Zero-order transmittance through 2DHAs in gold, as a function of the film thickness (P = 400 nm, a = 160 nm) **a** $\varepsilon_I = \varepsilon_{II} = 1.0$; $\varepsilon_{III} = 2.25$ **b** $\varepsilon_I = \varepsilon_{II} = \varepsilon_{III} = 2.25$. The spectral position as a function of w for both the EOT maximum (*triangular symbols*) an the EOT minimum (*circular symbols*) are shown in (**c**). *Dashed lines* summarize data obtained from (**a**), while *solid lines* are used for data taken from (**b**). The *horizontal dashed line* renders $\lambda_{SPP}^{(\pm1,0)}$. From Ref. [14]

400 nm (in order to obtain EOT in the visible). The metal is gold (with a frequency dependent dielectric constant, ε_m taken from Table 1.1). A schematic picture of the structure is shown as an inset in Fig. 2.11c.

Figure 2.11 renders the computed zero-order transmittance spectra through 2DHAs with different thicknesses, for both (a) asymmetric and (b) symmetric configurations.

Calculations have been conducted with the FDTD method. For optically thick films, we observe the "canonical" EOT resonant features appearing at wavelengths slightly red-shifted from the Rayleigh wavelength ($\lambda_R = \sqrt{\varepsilon_{III}}\, P = 600$ nm). As the film thickness is reduced, both maximum and minimum transmittance red-shifts by even hundreds of nanometers, while keeping high peak visibility.

In order to understand these spectral shifts, we analyze the EM modes bounded to the metal film. A flat unperforated optically thick metal layer supports a SPP on each surface. When the film thickness is reduced, these two modes interact, and are substantially coupled whenever the film thickness is smaller than 2–3 skin-depths. In this case, the dispersion relations of film modes can greatly differ from that of the SPP, while in the THz regime they remain close to the light line. We denote by $\vec{q}_{\text{mode}}(\lambda)$ the in-plane wavevector of these film modes (where the label "mode" can be either SPP, LR or SR) as a function of the wavelength λ. These film modes couple to external radiation and may lead to transmission resonances which, for small corrugations, are therefore expected to occur close to wavelengths satisfying:

$$\left(k_x^{in} + \frac{2\pi n}{P}\right)^2 + \left(k_y^{in} + \frac{2\pi m}{P}\right)^2 = q^2_{\text{mode}}(\lambda) \tag{2.3}$$

Here, the in-plane component of the incident wavevector is $\vec{k}^{in} = (k_x^{in}, k_y^{in})$, and n and m are integers. From now on, we denote by $\lambda_{\text{mode}}^{(n,m)}$ a wavelength that holds Eq. 2.3 at normal incidence ($\vec{k}^{in} = 0$), for some given values of n and m. Figure 2.11c shows the spectral positions of both minimum and maximum of the EOT peak appearing at largest wavelengths. We find that when the film is thick enough the EOT minimum very approximately coincides with $\lambda_{\text{SPP}}^{(\pm1,0)}$ [49]. In contrast, both maximum and minimum red-shift as film thickness reaches the "optically thin" regime.

To analyze whether the EOT phenomenon through optically thin 2DHAs has its origin in the excitation of an EM mode bounded to the film, we focus on the symmetric configuration with $w = 20$ nm. Figure 2.12a shows the transmission spectra for 2DHAs with different hole sizes. Vertical dashed lines mark different SR diffracted orders together with $\lambda_{\text{LR}}^{(\pm1,0)}$. The EOT spectral positions of both maximum and minimum wavelength approach $\lambda_{\text{SR}}^{(\pm1,0)}$ as the hole size decreases. At the same time, the EOT peak visibility is progressively reduced as the hole size decreases (the dashed line shows the result for the uniform film). Additionally, there are several small dips in the transmission spectra, which will be discussed later.

In order to assign even more conclusively EOT features to EM modes of the perforated film, we have calculated the band structure of surface modes in the holey film. The result is depicted in Fig. 2.12b (circular symbols), for a 2DHA with $a = 60$ nm. The dispersion relations for the bounded modes of a flat film (folded into the first Brillouin's zone), are represented with continuous lines. As usual, due to the presence of holes, the modes are coupled at the Brillouin's zone edges leading to band-gaps. In the wavelength window shown here, only the mode at the high-λ edge (labeled as $\widetilde{\lambda}_{\text{SR}}$) is related to an EOT peak at normal incidence, due to the structure symmetry [50, 51]. The dependence with hole size of $\widetilde{\lambda}_{\text{SR}}$, together with that of the spectral posi-

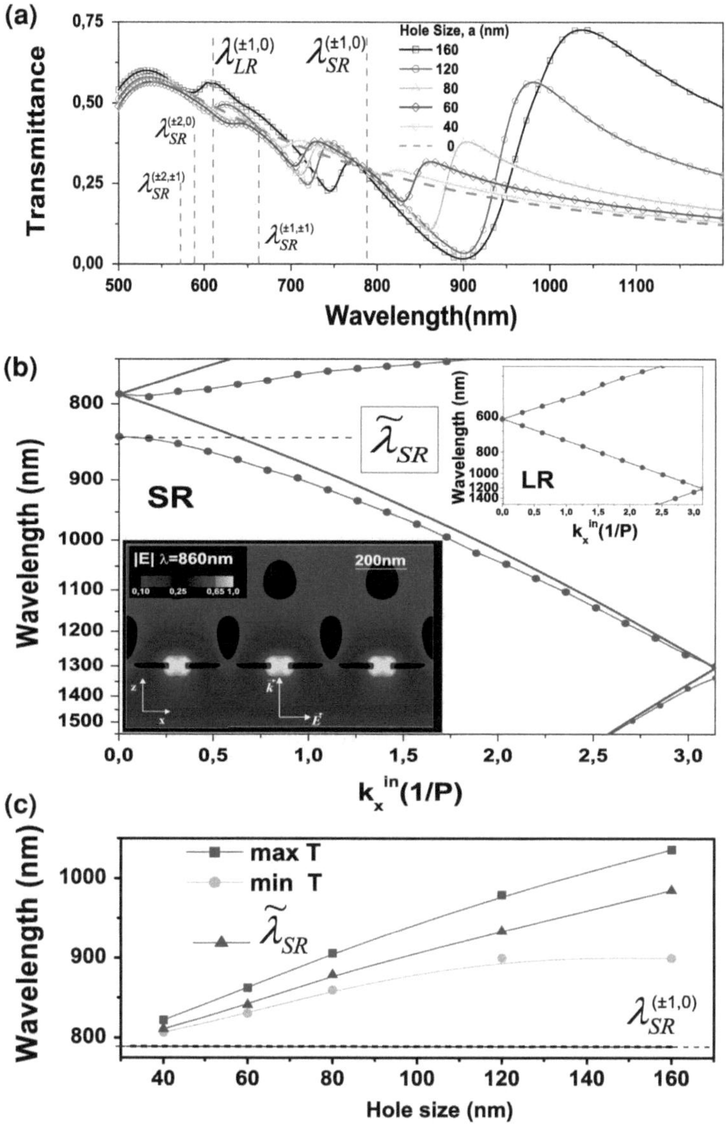

Fig. 2.12 For a holey thin film with $w = 20$ nm ($P = 400$ nm and $\varepsilon_I = \varepsilon_{II} = \varepsilon_{III} = 2.25$), **a** shows transmittance versus wavelength for different hole sizes. Vertical dashed lines display several values of $\lambda_{\mathrm{LR}}^{(n,m)}$ and $\lambda_{\mathrm{SR}}^{(n,m)}$ (see text) at $\vec{k^{in}} = 0$. **b** 2DHA dispersion relations along the x direction for $a = 60$ nm (*Circular symbols*). *Solid lines* represent the folded dispersion relations of LR and SR modes for the unperforated film. The *inset* in (**b**) shows a $|E|$ field map in the x-z plane ($y = P/2$) at the EOT wavelength. **c** EOT maxima (*square symbols*), minima (*circular symbols*) and $\widetilde{\lambda}_{\mathrm{SR}}$ (*triangular symbols*) as a function of the hole size. From Ref. [14]

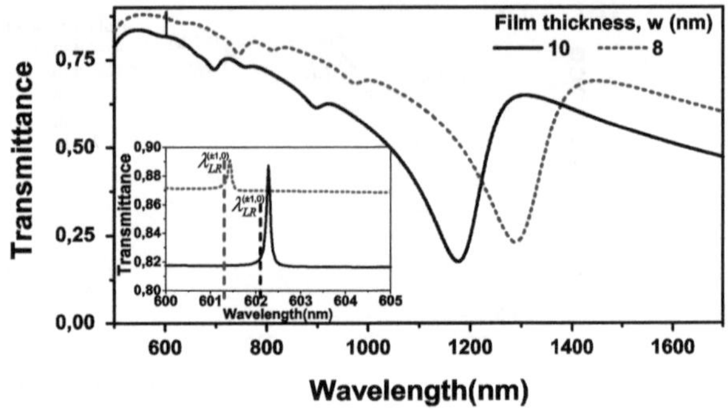

Fig. 2.13 Transmission spectra for two different film widths obtained with the approximate analytical method: $w = 10$ nm and $w = 8$ nm ($a = 160$ nm). *Inset* zoom close to the LR wavelengths (Corresponding $\lambda_{LR}^{(\pm1,0)}$ wavelengths are represented by *vertical dashed lines*). From Ref. [14]

tions of both maximum and minimum transmittance is displayed in Fig. 2.12c. For each hole size $\widetilde{\lambda}_{SR}$ lies between the spectral positions of the transmission maximum and minimum. Nevertheless, as the hole size shrinks to zero, the minimum of transmittance tends to $\widetilde{\lambda}_{SR}$. The inset of Fig. 2.12b renders a $|E|$ field map at the EOT peak wavelength, showing that the field enhancement around the holes [17, 37] is also present in optically thin films.

Interestingly, LRs do not noticeably contribute to transmission in the FDTD calculations (Fig. 2.12a). Notice that, due to the antisymmetric charge distribution of the LR, its field is almost negligible inside metal and it is less bounded to the surface than a SR mode. Therefore, the LR is both less absorbed and worse coupled to radiation than the SR. In short, the LR field is perturbed very weakly by the holes, so the coupling with the incident light diminishes. A consequence of this, is that the LR band structure for the drilled film virtually coincides with the unperforated one (Inset Fig. 2.12b). This weak coupling to radiation modes implies long times of the EM field standing at the surface. This suggest that the LR resonance could have been missed given the finite simulation time available. To be sure that LRs are not related to the shallow transmission dips, we have developed an approximate method for solving Maxwell's equations. In this method, the field is represented as a Fourier-Floquet series in the x-y plane and a power series in the coordinate perpendicular to the layer, z [51]. This approach works only for extremely thin metal films (thinner than what is experimentally achievable in a continuous film nowadays), so it has mainly academic value. Nevertheless, it is useful for understanding the underlying physics. Figure 2.13 renders transmission spectra calculated with the approximate method for a 8 nm and 10 nm thin films. The zoom in wavelengths close to $\lambda_{\mathrm{LR}}^{(\pm1,0)}$ (inset to Fig. 2.13) reveals that extremely narrow peaks can be associated to LR modes. Anyway, spectral resolution within the FDTD method does not allow LR peaks to be resolved. The detection of this transmittance peaks due to LR plasmons would be even more difficult from the experimental point of view due to the finite size of the samples. In any case, this analysis shows that the small dips found with the FDTD method at short wavelengths are exclusively related to higher SR diffracted orders.

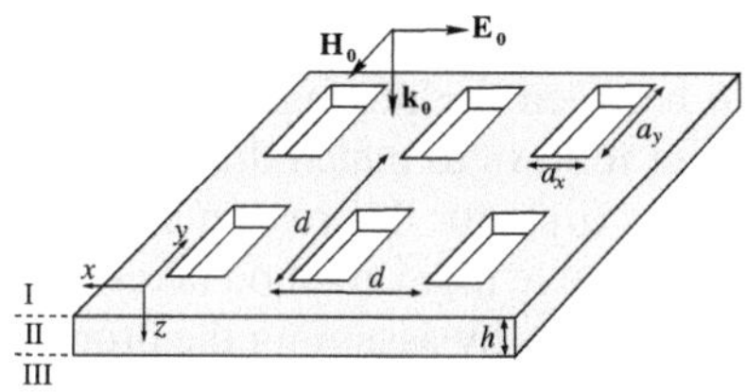

Fig. 2.14 Schematic picture of a square array of rectangular holes of side a_x and a_y perforated on a free-standing silver film of thickness h. Parameter d defines the period of the array. The apertures are illuminated by a p-polarized plane wave at normal incidence

2.4 The Role of Hole Shape on EOT Through Arrays of Rectangular Holes

Several works shown that the shape of the subwavelength aperture also has a great importance in the transmission spectra [27–29, 52, 53]. These studies demonstrated that elliptical or rectangular holes dramatically influence not only the resonant wavelength but also the polarization and the final transmittance. On the other hand, several experimental and theoretical works have shown that even a single rectangular hole exhibits a localized transmission resonance emerging at around the cutoff wavelength, λ_c, of the hole waveguide [30, 31, 42, 54]. This resonance can be understood as a Fabry-Perot resonance in which the propagation constant inside the hole is zero.

The aim of this section is to analyze in detail the interplay between these two types of transmission resonances (SPP and cut-off resonance) that are operating in a 2D array of rectangular holes. Figure 2.14 shows schematically the system under study: an infinite array of rectangular apertures of sides a_x and a_y perforated on a free-standing silver film of thickness h. The structure is illuminated by a p-polarized plane wave at normal incidence (i.e. incoming electric field is pointing along the short edge of the holes). In our calculations, we use the same geometrical parameters as those of the experiments carried out in Ref. [29]: $a_x = 200$ nm, $a_y = 260$ nm and $h = 400$ nm. The period of the array, d, will be varied between 500 and 900 nm.

Figure 2.15b shows the normalized-to-area transmittance spectra corresponding to the geometrical parameters of Ref. [29], calculated with the couple mode method (CMM) (Sect. 1.3). To treat the metal properly, surface-impedance boundary conditions (SIBCs) are imposed on the metallic boundaries, except on the vertical walls of the holes which are treated as perfect conductor surfaces. To consider only the fundamental TE eigenmode (TE_{01}) in the modal expansion within the holes gives accurate results for the transmittance spectrum. As it was stated in Sect. 1.3, in order to improve the accuracy of the method realistic values for the propagation constant of the fundamental mode, q_z, are incorporated into the formalism by using the effective index method [55]. Note that the electric field of the TE_{01} mode points along the x-direction, i.e., it is parallel to the short side of holes (a_x) and perpendicular to a_y. The properties of the two parallel real metal plates TM mode are the starting point of the method. From its characteristic equation, it is obtained an effective dielectric constant (ε_d) as the ratio between the propagation constant of that mode and the vacuum one. This procedure follows to match the boundary conditions of the electric field across the long sides (a_y), for this reason the metal surfaces are separated by a distance a_x. Next, the effective index method takes into account the penetration of the electric field at the short axis by assuming the propagation constant of the TE mode (in the same system), but being in this case a_y the distance between surface faces. Finally, the value of q_z is obtained after solving the characteristic equation of such mode where the dielectric constant of the region between surfaces is set to ε_d. Note that by analyzing the location in which q_z changes from being a real quantity to a purely imaginary magnitude, we can calculate λ_c within this approach. For the particular set of geometrical parameters of the holes forming the 2D array here chosen, $\lambda_c = 695$ nm. Different curves correspond to different periods of the 2D square array (ranging from $d = 500$ to $d = 900$ nm). To compare them with the FDTD method results (Sect. 1.2), Fig. 2.15a depicts the corresponding transmission spectra obtained. It is clear that CMM is able to capture accurately the main features observed in the FDTD spectra. The locations and linewidths of the several peaks are well reproduced within the CMM approach. The heights of the transmission peaks are higher in CMM, mainly due to the fact that absorption within the vertical walls of the holes is neglected. It is also important to note that the theoretical results of panel (b) are in very good agreement with the experimental data [29].

Now we concentrate on analyzing the physical origin of the different transmission peaks appearing in Fig. 2.15b by just studying this simple formula for transmission (See Sect. 1.3):

$$T = \frac{|I_o|^2}{4\sqrt{\varepsilon}} \frac{G_i}{(G_i)^2 + \left(\frac{|G-\Sigma|^2 - |G_v|^2}{2|G_v|}\right)^2} \tag{2.4}$$

For rectangular holes and within the SIBC approximation the different terms in Eq. 2.4 are:

The term I_0 measures the overlap between the incident plane wave and the 0-mode inside the hole:

$$I_0 = \frac{\imath\, 4\sqrt{2}}{\pi(1+Z_s)} \sqrt{\frac{a_x a_y}{d_x d_y}} \tag{2.5}$$

The term G_v controls the coupling between the input and output sides of the holes:

$$G_v = \frac{2\, i\, Y_0\, e^{i q_z h}}{e^{2 i q_z h}(1+Z_s Y_0)^2 - (1-Z_s Y_0)^2} \tag{2.6}$$

The expression for the *self-energy* Σ is given by:

$$\Sigma = i\, Y_0 \frac{e^{2 i q_z h}(1+Z_s Y_0) + (1-Z_s Y_0)}{e^{2 i q_z h}(1+Z_s Y_0)^2 - (1-Z_s Y_0)^2} \tag{2.7}$$

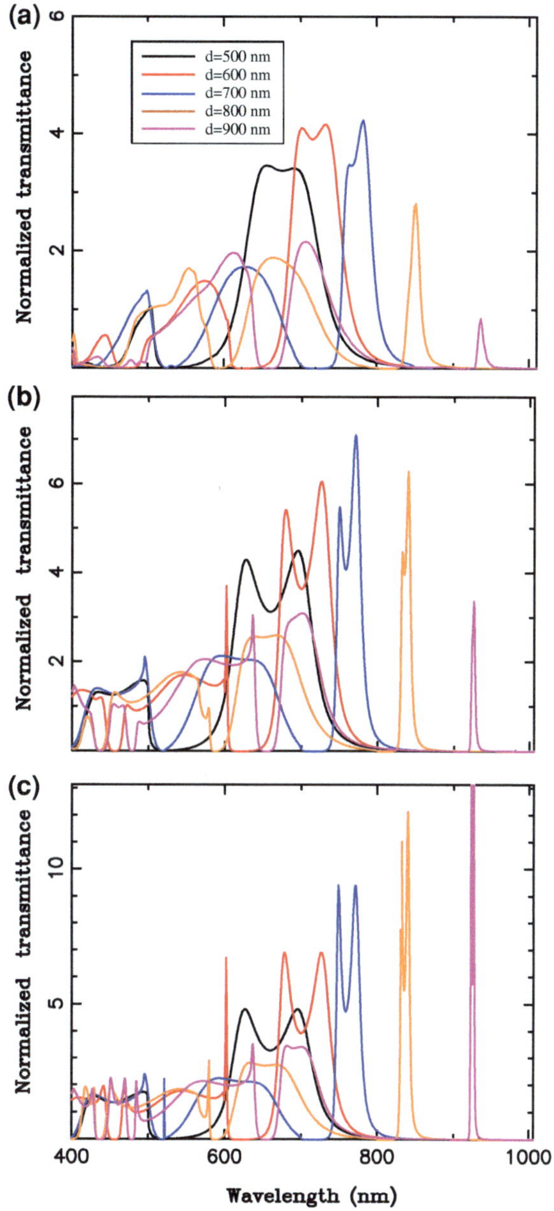

Fig. 2.15
a Normalized-to-area transmittances calculated with the FDTD method for different values of the lattice period d. **b** and **c** show the normalized-to-area transmittances calculated with the CMM. In (**b**), the dielectric function is that of silver whereas in (**c**) absorption of silver is neglected ($Im\,\varepsilon(\lambda) = 0$)

Finally, the EM-coupling between the holes forming the 2D array is mediated by the term G, which can be expressed as:

$$G = \frac{i\,a_x\,a_y}{2\,d_x\,d_y} \sum_{l=-\infty}^{l=+\infty} \sum_{m=-\infty}^{+\infty} \frac{k_0\,(k_0 + Z_s k_z) - k_m^2}{(k_z + Z_s k_0)\,(k_0 + Z_s k_z)}$$
$$\times\, \mathrm{sinc}^2\left(\frac{k_l\,a_x}{2}\right)\left[\mathrm{sinc}\left(\frac{k_m a_y + \pi}{2}\right) + \mathrm{sinc}\left(\frac{k_m a_y - \pi}{2}\right)\right]^2 \quad (2.8)$$

with $k_l = \frac{2\pi}{d_x} l$, $k_m = \frac{2\pi}{d_y} m$, $k_p = \sqrt{k_l^2 + k_m^2}$ and $k_z = \sqrt{k_0^2 - k_p^2}$.

Here $Z_s = 1/\sqrt{\varepsilon_m}$ (being ε_m the dielectric constant of the metal), and ε the dielectric constant of the surrounding media (in this case vacuum). On the other hand, $Y_0 = q_z/k_0$ corresponds to the admittance of the fundamental mode with $k_0 = 2\pi/\lambda$.

It can be demonstrated from Eq. 2.4 that transmission maxima and minima coincide with solutions of $|G - \Sigma| = |G_v|$, for 2DHAs at $\lambda > \sqrt{\varepsilon}\,d$. In other words, as in a typical resonant phenomenon the transmittance is governed by the behavior of the denominator. In Fig. 2.16 we show the dependence of both $|G - \Sigma|$ and $|G_v|$ versus wavelength for two limiting values of *d*, $d = 500$ nm and $d = 800$ nm (notice that G_v does not depend on *d*). Interestingly, the spectral locations of the transmission peaks in Fig. 2.15b are marked by the cuts between $|G - \Sigma|$ and $|G_v|$. It is worth comparing the behavior of $|G - \Sigma|$ for the periodic arrays with $d = 500$, 800 nm with the corresponding $|G - \Sigma|$ for a single rectangular hole (dashed line in Fig. 2.16). The transmittance through a single rectangular hole is also governed by Eq. 2.4 but with a different G-term (the sum over diffractive modes in Eq. 2.8 is replaced by an integral over the continuous spectrum of plane waves). For $d = 500$ nm, $|G - \Sigma|$ (full black line) at the wavelength region of interest (near $\lambda = 700$ nm) is close to the single hole counterpart. It is expected then that the nature of the transmission resonances will be similar for a 2D array and for an isolated rectangular hole. However, there is a difference between the single hole case and the 2D array for this value of *d*. Whereas in the 2D array, $|G - \Sigma| = |G_v|$ at two different wavelengths (leading to two transmission peaks), for the single hole there is only one transmission peak appearing at a wavelength in which the difference between $|G - \Sigma|$ and $|G_v|$ is minimal. The reason of this distinct behavior stems from the Fabry-Perot nature of this transmission resonance. In a Fabry–Perot cavity, the spectral locations of the transmission resonances strongly depend on the reflectivity at the edges of the cavity. Our results indicate that the presence of a $2D$ array modifies the reflectivity of the metallic interface when compared to the single hole case, leading to the appearance of two transmission peaks. For $d = 800$ nm (blue line), $|G - \Sigma|$ present additional features located at $\lambda \approx 600$ nm and $\lambda \approx 800$ nm. These correspond to zeroes of the denominator of G (see Eq. 2.8), appearing at the condition $k_z + Z_s k_0 = 0$. This condition is nothing else than the equation for the excitation of a SPP on a non-corrugated (no holes) surface of a metal film within the SIBC approach. Note that the cuts between $|G - \Sigma|$ and $|G_v|$ appears at wavelengths slightly larger than this condition. Therefore, the character of the two transmission peaks for $d = 800$ nm emerging at $\lambda \approx 830$ nm will be quite similar to a SPP. The two transmission peaks are associated with the symmetric and the anti-symmetric combinations of the two SPPs

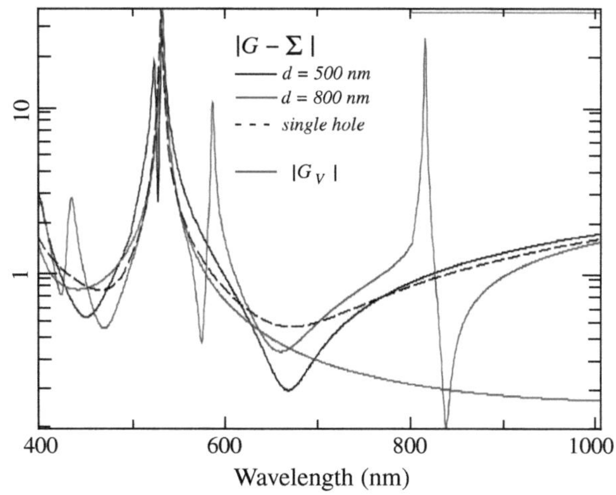

Fig. 2.16 Absolute values of $|G - \Sigma|$ and $|G_v|$ terms (in logarithmic scale) for two different lattice constants of the hole array ($d = 500$ nm and $d = 800$ nm) and also for a single hole

at the two surfaces which are evanescently coupled through the holes, as explained in Ref. [4].

As a general conclusion about the nature of the transmission resonances appearing in 2D arrays, we could state that the two mechanisms leading to EOT (localized resonance and SPP-based) cannot be simply separated as done in previous studies [54] [see comment by Cheng-ping Huang and Yong-yuan Zhu, arXiv:0706.0250v1 (unpublished)] and [56]. Even for the shortest and the longest period considered here, the transmission resonances benefit from both mechanisms. Therefore, these transmission resonances have a hybrid character. Hybrid resonances have already been observed and studied previously in lamellar [57, 58] and bottle-shaped gratings [59]

The physical picture described above is reinforced when looking at the evolution of the electric field patterns associated with the resonant process as a function of d. In Fig. 2.17 we plot the E-field amplitude at resonance (normalized to that of the incident plane wave) evaluated at the $z = 0^{-}$ interface. Four periods are considered here: $d = 600, 700, 800$ and 900 nm. The electric field amplitude for $d = 600$ nm is mainly concentrated over the holes as corresponds to the excitation of a localized resonance (cut-off resonance) inside the holes. As d is increased, the character of the resonance changes gradually. For $d = 800, 900$ nm, the E-field intensity maxima are along the ridges of the holes as corresponds to a SPP wave propagating in the x-direction. For $d = 700$ nm, there is a mixing between the two mechanisms as fingerprints of the SPP wave begins to emerge at the edges of the holes. The near field distribution exhibits an intermediate character between the two limiting behaviors (SPP and cut-off resonances).

Finally, we would like to address the question of why there is a kind of optimum value for d when looking at the evolution of the transmission peaks with the period of the array (see Fig. 2.15a,b). In our calculations this optimum d is around 650–700 nm, close to the cut-off wavelength, 695 nm. Naively, this could imply that the optimum

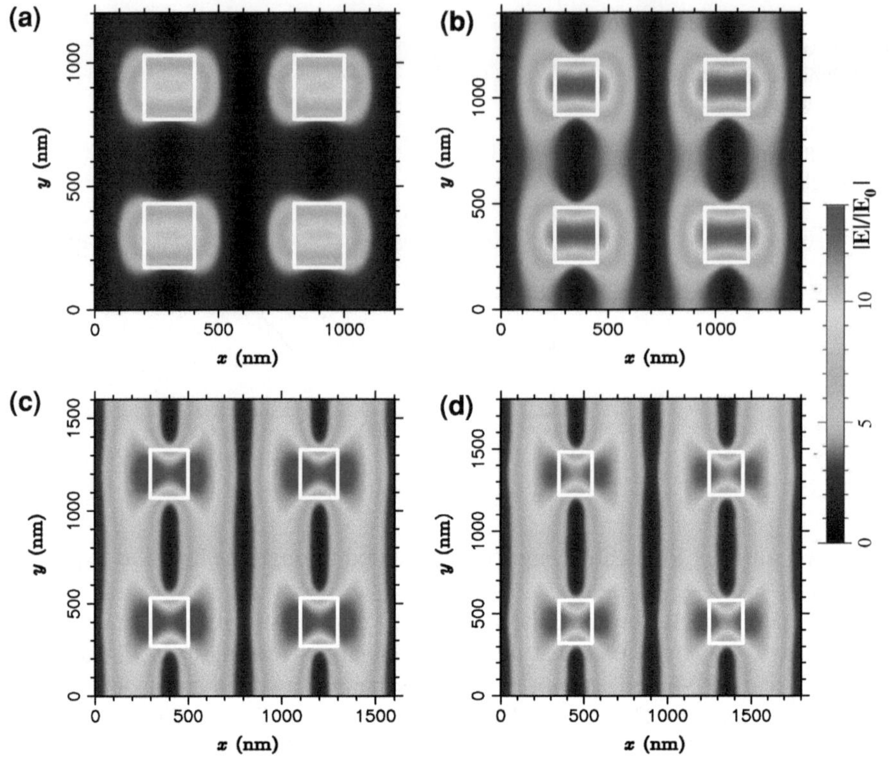

Fig. 2.17 Electric field amplitude calculated at the resonant wavelength and evaluated at the $z = 0^-$ plane for **a** $d = 600$ nm, **b** $d = 700$ nm, **c** $d = 800$ nm and **d** $d = 900$ nm. The white lines mark the positions of the holes. The E-field magnitude is normalized to the incident one

d appears when the resonant wavelengths of the two mechanisms (SPP and cut-off resonance) coincide. However, panel (c) of Fig. 2.15 demonstrates that the explanation is more complex. If the absorption in the metal is neglected, the heights of the transmission peaks grow with d like d^2, as it would correspond to a perfect transmission (100%) per unit cell [60]. Therefore, EOT associated with SPP excitation is only limited by absorption. As explained above, absorption along the walls of the holes is not taken into account in the approximated model. However, we have checked that for all d analyzed in this study, E-field intensity maxima are located at the horizontal metallic surfaces, where SIBCs are imposed within the modeling. Therefore, considering only absorption on top and bottom surfaces of the metallic film is a reasonable approximation when analyzing the evolution of the heights of the transmission peaks as a function of d.

Within this approach, it is worth defining two different lifetimes operating during the transmission process. By looking at the linewidth of the transmission peaks with no absorption (Fig. 2.15c), we can extract the lifetime associated with the resonant process, τ_{res}. This quantity is related with the radiation losses as a result of the

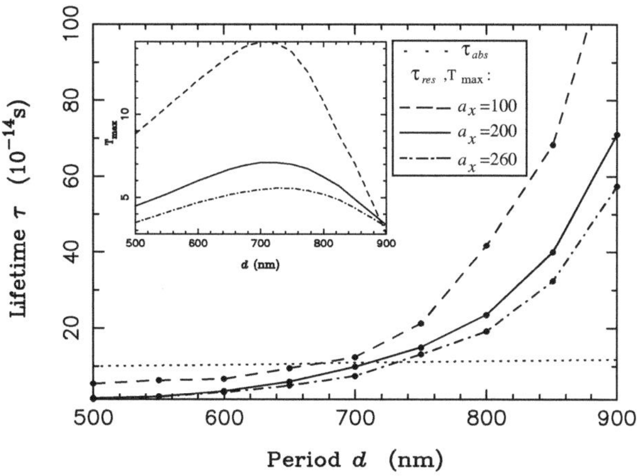

Fig. 2.18 Absorption lifetime τ_{abs} (*dotted line*) and lifetime of the resonant process τ_{res} versus period of the hole array for $a_x =$ 100 nm (*dashed line*), $a_x = 200$ nm (*solid line*) and $a_x = 260$ nm (*dot-dashed line*). *Inset* Transmittance calculated at resonance versus *d*

coupling of holes to radiation. It is depicted in Fig. 2.18 as a function of the period of the array for three values of a_x (200, 100 and 260 nm). As expected, when *d* is increased, resonant lifetime grows rapidly. On the other hand, absorption introduces another time into the problem. From the knowledge of $\varepsilon(\lambda)$, we can estimate the time taken for a photon to get absorbed, τ_{abs}. This lifetime is almost independent on λ, as shown in Fig. 2.18 (dotted line). It is expected that optimum *d* would appear where $\tau_{\text{res}}(d) \approx \tau_{\text{abs}}(d)$. The line of reasoning leading to this naive rule is the following. When $\tau_{\text{res}}(d)$ is much smaller than $\tau_{\text{abs}}(d)$, photons are mainly transmitted and they are not absorbed by the metal. Absorption plays a minor role in the transmission process and the normalized-to-area transmittance at resonance increases when *d* is increased, as seen in Fig. 2.15c. In the other limit ($\tau_{\text{abs}}(d)$ being much smaller than $\tau_{\text{res}}(d)$), photons are absorbed by the system before the resonance is built up. As $\tau_{\text{res}}(d)$ grows quadratically with *d*, a decrease of the transmittance at resonance versus *d* is expected to occur in this limit. When interpolating between these two limits, it is clear that the curve displaying the transmittance at resonance versus *d* should present a maximum for an *optimum* value of *d*. As $\tau_{\text{res}}(d)$ evolves very rapidly with *d* whereas $\tau_{\text{abs}}(d)$ is almost independent on *d*, optimum *d* should appear close to the condition $\tau_{\text{abs}}(d) = \tau_{\text{res}}(d)$. Figure 2.18 demonstrates that this last condition marks the location of the optimum *d* for the three different values of a_x analyzed. Therefore, we can safely conclude that the physical origin of the optimum *d* observed in calculations stems from the absorption present in the metallic film.

Note that in the experiments [29], the finite size of the hole array introduces a third lifetime associated with the spatial extension of the array, τ_{size} [41]. As the absorption in the metal, it also acts as a limiting factor in the final transmittance. If the number of holes is large enough, τ_{size} is greater than τ_{abs} and then the limiting factor would be the absorption by the metal, as in the case of an infinite array. However, if the array is very small, τ_{size} would be smaller than τ_{abs} and finite size effects would control the optimum *d*. Then the cut between $\tau_{\text{res}}(d)$ and τ_{size} would mark the location of optimum *d*. If this is the case ($\tau_{\text{size}} < \tau_{\text{abs}}$), then the cut would appear at a shorter *d*.

This seems to be the case in the experiments reported in Ref. [29] as the experimental optimum d is of the order of 600 nm, instead of a value close to 700 nm obtained from our calculations for an infinite array of holes (see Fig. 2.18).

2.5 EOT Through Metal-Coated Monolayers of Microspheres

Several papers on EOT involve experiments and simulations of metal films and gratings of rather simple geometry [2, 4, 8, 28, 61–66]. One drawback, considering applications, with these structures is that they are usually produced using a material removal focused ion beam and/or complex lithographic methods. In this section we analyze a quite different structure, namely, slabs of self-assambled arrays of dielectric microspheres covered with thin metallic layers. As we will see, these kind of systems present many similarities in their transmission properties with those in perforated metal films, with the advantage of being easily deposited over relatively large areas. Moreover, the following metal deposition can be done by standard techniques.

Monolayers of close-packed arrays of microspheres behave like two dimensional photonic crystal slabs (PCS) with photonic modes that may couple to the incident light. It has been shown that the strength of coupling and the position of the observed transmission dips could easily be altered by deposition of, e.g., amorphous Si onto the microsphere arrays [67, 68]. If the microsphere array is instead covered with a thin metal film, EOT through the slab has been observed [69]. Different mechanisms could be responsible for the EOT-like transmission features: Mie resonances of the spheres, transmission through the empty spaces left by the spheres after covering them with the metal,... [69]. In order to clarify the actual origin of the resonances observed, in this section we analyze a set of experimental results carried out in the group led by Prof. D. Bäuerle, from the Institute of Applied Physics Johannes-Kepler-Universitat in Linz (Austria). For this, we use the FDTD method discussed in Sect. 1.2. The good agreement between measured and modeled spectra allows further in-depth interpretation of the origin of the different features observed in the measured transmission spectra, highlighting the relevance of waveguide modes in the microsphere array on the EOT properties.

2.5.1 Methods

Experiment

Microspheres of different materials have been utilized in various fields of research in the past few years. Examples of applications are micro-resonators with high quality factors [71], in mask lithography [70], and also as lens arrays for different types of laser-induced micro- and nano-patterning of material surfaces. In the case of laser-induced applications, close-packed 2D-lattices of usually transparent microspheres

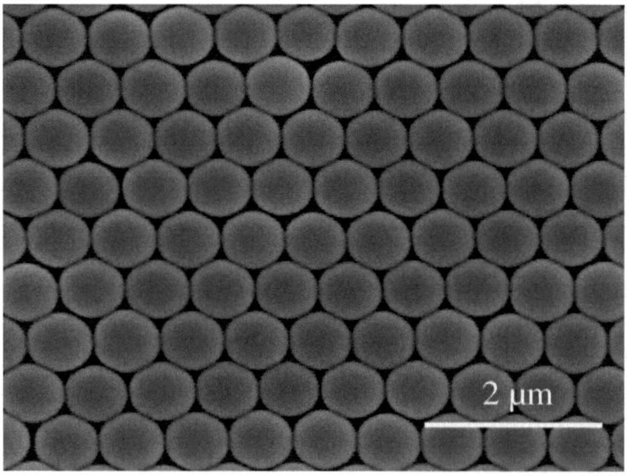

Fig. 2.19 Scanning electron microscope (SEM) picture of a Ni-coated monolayer of quartz ($a - SiO_2$) microspheres of diameter $d = 0.72\,\mu$m. The support is a 1.0 mm thick $a - SiO_2$ platelet. From Ref. [16]

are used as a lens array allowing single step large area parallel processing [72–76]. Among those are patterns generated from metal-coated monolayers of microspheres by laser-induced forward transfer (LIFT) [77–80].

Through the experiments, close-packed monolayers of amorphous silica ($a - SiO_2$) or polystyrene (PS) microspheres (diameters $d = 0.39, 0.78, 1.0,$ and $1.42\,\mu$m) were deposited on quartz supports (1 mm thick) using colloidal suspensions. The monolayers were covered with different metals (Ag, Au, Ni) and film thickness (30–300 nm) using standard evaporation techniques. A typical metal covered monolayer is shown in Fig. 2.19. The metal films cover approximately the upper half of single spheres, while the lower half remains uncoated. At the top of spheres the thickness of the coating (75 nm) is about equal to that measured with a nearby quartz crystal microbalance (QCM). Towards the edge of spheres the film thickness slightly decreases. In the interstices between the spheres, the coating is placed on the quartz support. Within these areas, the film thickness measured by means of an atomic force microscope (AFM), is equal to that measured by QCM. The areas of close-packed monolayers were, typically, of the order of $\sim$cm^2. Because of certain size dispersion of the microspheres and the deposition technique employed [81], the monolayers exhibit a polycrystalline structure with a typical domain size of about 50–100 m . Transmission experiments were performed at normal incidence both on bare PCSs and on the covered with metal slabs. The transmission measurements were done in the far-field, in a configuration that only collected the zero-order transmission. Since aperture diameters of 1–3 mm were used for the transmission measurements, any polarization dependent effects could not be probed and non-polarized light was used.

FDTD Modeling

Simulations were performed by using the FDTD method. A small grid size of 6 nm was used in all reported results. The dielectric constant of the different metals considered were taken from their bulk values, and approximated by a Drude–Lorentz functional form (Table 1.1). Dielectric constants for the quartz support, and the silica and polystyrene microspheres were assumed wavelength independent and set to 1.52, 1.392 and 1.572, respectively. As the geometry of the metal layer is not precisely known, for simplicity the thickness of the metal film on the top of each sphere was assumed to be constant (72 nm). We expect that this simplification of the metal geometry will induce at most some small spectral displacements of the transmission resonances and of the average transmittance, but will otherwise have a negligible effect on the overall transmission properties of the system. In order to compare with the experimental transmittance, only transmission into the zeroth diffraction order was computed.

2.5.2 Results and Discussion

Figure 2.20 compares measured and modeled spectra for silica sphere arrays covered with Ni, Ag and Au metals. Overall, the modeled spectra reproduce quite well the observed features with respect to both the absolute transmission values and peak positions. The main difference that can be observed is the additional peak at around 1300 nm in the calculated spectra. For the case of Ni, this peak seems to be hidden under the shoulder of the main peak. For Ag and Au it seems to be absent in the measured spectra. All peaks in the measured spectra are also slightly broader, likely because of the size-dispersion of the spheres and the polycrystalline structure of the array. Nevertheless, the overall good agreement suggests that this simple fabrication route leads to structures where disorder is small enough as not to spoil the optical transmission resonances expected in perfect arrays.

In the optical regime, the different behavior of the metals is notable. The peaks for Ni are less pronounced, while Ag and Au presents more detailed fine structure in both measured and modeled spectra. This effect is related to the difference in optical properties of these metals. As discussed in Sect. 2.2, Nickel is a less "ideal" metal than Ag and Au, with relatively high absorption in the wavelength region of interest, resulting in less pronounced features in both measured and modeled spectra. The calculated spectra for gold and silver have more defined (and stronger) peaks in comparison to the measured ones at shorter wavelengths. This may be due to the presence of disorder in the sample, where not all unit cells are strictly equal. Also, films deposited onto the microsphere arrays exhibit a poly/nano-crystalline structure, which may alter the optical properties of the metal relative to bulk values used in the calculations. Both such alterations impair resonant behavior, being therefore more evident in Ag and Au than in Ni (where resonances are already hampered by intrinsic absorption of the metal).

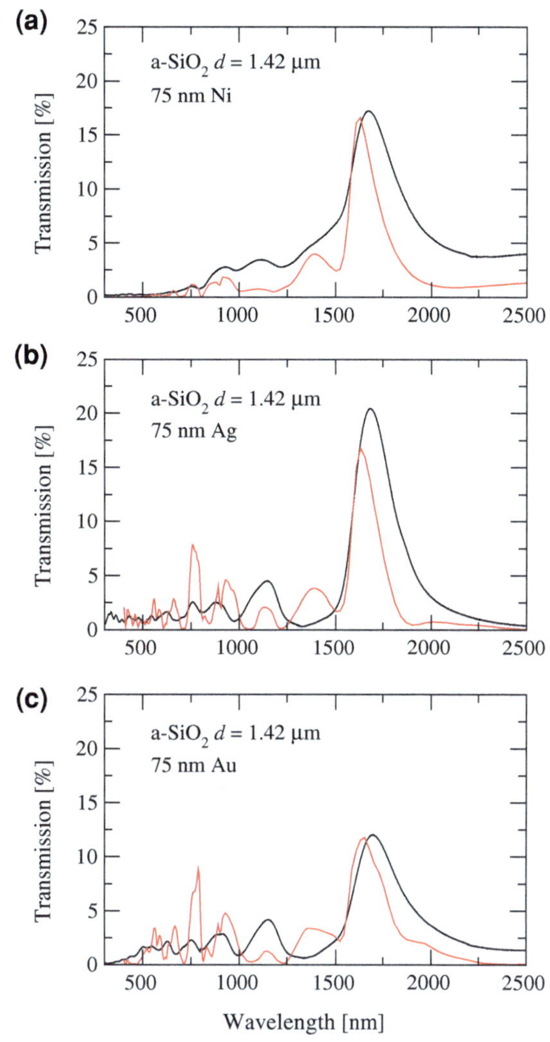

Fig. 2.20 Measured (*black*) and calculated (*red*) zero order transmission through metal coated MLs of $a-SiO_2$ microspheres ($d = 1.42\,\mu$m). From Ref. [16]

The situation is different in the telecom regime (transmission peaks appearing around 1600 nm in Fig. 2.20). In this case, the transmission level for all metals considered is similar, being even larger for Ni than for Au or Ag. Notice that, in this case, the full-width-at-half-maximum ($\Delta\lambda$) is very very similar for all three metals considered: 163 nm for Ni, 154 nm for Ag and 126 nm for Au. Given that the dielectric constant of Ni is very different from that of Ag and Au, this implies that the time that the electromagnetic field stays at the structure is limited by radiation, more than by absorption. This time can be estimated as $T = \lambda_{\max}^2/(c\Delta\lambda)$ and the distance that the EM field travels on the surface as $L_T = \lambda_{\max}^2/\Delta\lambda$, where $\lambda_{\max}$ is the spectral position of the transmission maximum and c the speed of light. From the simulation we can estimate $L_T = 16.2\,\mu\text{m}$, $17.4\,\mu\text{m}$ and $12.6\,\mu\text{m}$ for Ni, Ag and Au, respectively. This values are smaller than, for instance, the propagation lengths of surface plasmon polaritons at $\lambda = 1600$ nm (which approximately are 25 μm for Ni, 360 μm for Ag and 360 μm for Au), which reenforces the hypothesis that radiation losses dominate over absorption. Notice also that the larger absorption in the case of Ni could be compensated by the larger skin depth (33 nm in Ni, 22 nm in Ag and 23 nm in Au for $\lambda \approx 1600$ nm), which implies both a larger direct transmission through the metal layer and a larger effective hole radius.

To further study the behavior of this composite structure and the validity of using the FDTD method, different parameters were investigated. Here, the refractive index of the spheres was changed by considering polystyrene microspheres. The sphere diameters (periodicity) was also altered, see Fig. 2.21. As expected, by using monolayers of polystyrene spheres (with a higher refractive index than $a - SiO_2$) with different diameters, one finds that the main peak shifts with the periodicity of the array. Again, measured and modeled spectra show good agreement (Fig. 2.21). We associate the higher values for the calculated peaks both to disorder in the actual sample and to the fact that absorption in the PS spheres was neglected in the calculations.

The main peak is further red-shifted relative to the diameter by about a factor of 1.3 *d*, whereas a factor of 1.2 *d* was observed for the silica spheres (Fig. 2.20). This is related to the higher refractive index of the polystyrene spheres relative to silica. The same effect is observed for PCS without metal, that is, a higher "effective" refractive index red-shifts the main minima (dip) in transmission [67, 68]. Additionally, the main dip in the dotted curves in Fig. 2.21, that show the transmission of the bare MLs, and the main transmission peak of the metal coated arrays show a clear correlation. The main transmission peak is slightly red-shifted compared to the main dip. This behavior implies that the transmission is related to the supported modes of the bare (uncoated) 2D-PCS as suggested earlier [69]. The transmission spectra of the bare PCSs are also included in the graphs where, once again, the differences in the transmission curves can be seen. The modeled spectra show much narrower main dips than the measured ones, again pointing to the influence of absorption in the PS spheres, and also to size dispersion of spheres and grain boundaries within the monolayers. In any case, the fabrication method allows for simple scaling (positioning) of any transmission peak (or dip) of interest, as can be seen in Fig. 2.21. It is also demonstrated in Fig. 2.21c that the main peak can be easily shifted to the visible wavelength region. This could be interesting with possible application for these composite structures as, e.g., for the fabrication of relatively narrow band filters.

Importantly, the close spectral correspondence between transmission dips in the uncoated system and transmission peaks in the coated one is also present in the calculation, even more clearly so, as spectral features are narrower here than in the experiment. The thickness of the metal deposit was also varied for both Ag and Au metals on silica spheres, see Fig. 2.22. For both metals, the intensity of the main transmission peak decreases roughly exponentially. Notice that the measured transmission is higher for thicker deposits compared to modeled spectra, which we associate to the assumption of homogenous film thickness in the calculations. Again, quite large

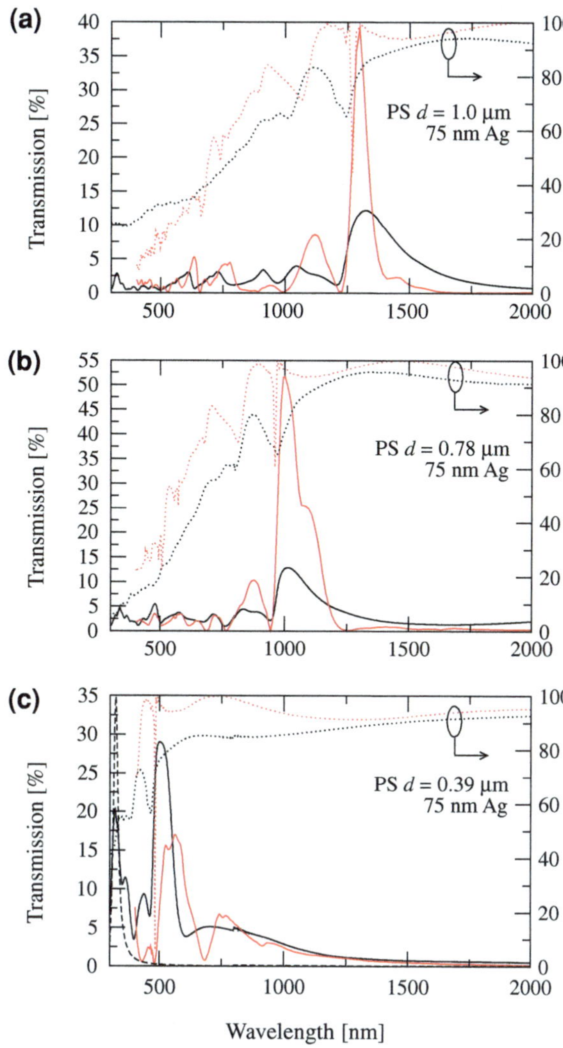

Fig. 2.21 Measured (*black*) and modeled (*red*) zero order transmission through monolayers of polystyrene spheres. Bare monolayer-dotted curves (*right* y-scales) and coated with 75 nm silver-full curves (*left* y-scales). **a** $d = 1.0\,\mu$m, **b** $d = 0.78\,\mu$m, **c** $d = 0.39\,\mu$m. The main peak (or dip) scales with the periodicity (**d**). From Ref. [16]

discrepancies can be observed between measured and calculated spectra in the short wavelength region. Also, the main peak red shifts as the thickness is increased (More pronounced for the measured spectra). Possibly, this can be related to coupling of the modes on the two interfaces; PCS/metal and metal/air. For films with thickness less than 50 nm, two peaks can be observed in the calculated spectra, whereas only one peak is observed for the thicker deposits, suggesting a coupling/decoupling behavior of the two modes as the thickness is increased.

The rest of the section is devoted to ascertain which are the relevant mechanisms for the transmission resonances in this system. Notice that the composite slab is quite complex, and transmission resonances could be due to one or several factors,

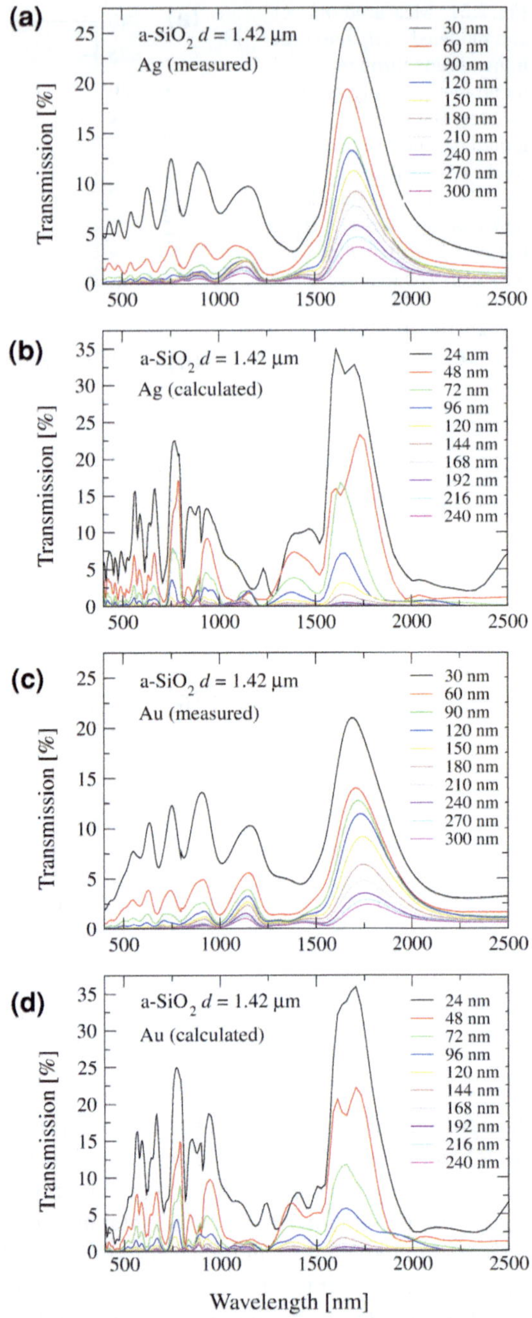

Fig. 2.22 Measured (**a**, **c**) and calculated (**b**, **d**) transmission spectra for different thicknesses of the metal deposit. Ag and Au were used on $d = 1.42\,\mu$m silica spheres. From Ref. [16].

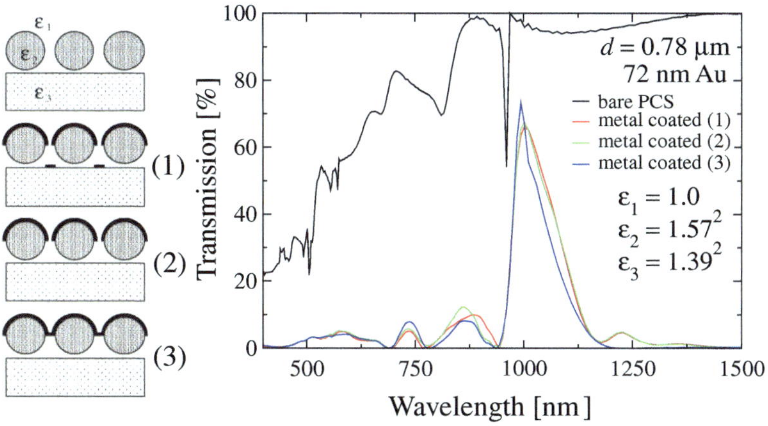

Fig. 2.23 Calculated transmission curves for bare PCS and with different metal coverage (1)–(3). From Ref. [16].

like: surface plasmons coupled either through the holes in the interstices or through the metal (if the metal film is optically thin), photonic crystal modes in the sphere layer (weakly or strongly coupled), Mie resonances of the spheres, particle plasmon modes of the triangular metal deposit on the quartz support, etc. The good agreement between measured and calculated spectra allows us to study of the relevance of these different mechanisms, through the modeling of similar but simpler systems.

To start with, two different but related structures were modeled: Metal coated sphere arrays without any metal on the support and sphere arrays fully covered with metal (and consequently no metal on the support either). The calculated results are shown in Fig. 2.23. Interestingly, the calculations reveal that the metal deposit on the support has negligible influence on the overall transmission and, more importantly, that the transmission spectra remains practically unaltered if the holey metal cap covering the spheres is replaced by a continuous metal cap. So, for this parameter range, the coupling across the metal film is mainly due to coupling through the metal, and not through the holes. This calculation also shows that there is no need for improvements in the fabrication process in order to get rid of the deposited metal particles in the substrate. In order to investigate the importance of a PCS and its guided modes as support to the metal film, the spheres were simply removed in the model system by introducing a uniform refractive index below the corrugated metal film (both with and without holes). In this case, the transmission process can be explained by a resonant model involving surface plasmon excitations and tunneling through the corrugated thin metal film [4, 37]. The results are depicted in Fig. 2.24. Remarkably, in the uniform dielectric case, absolute transmission values are much lower than those obtained for the sphere system. In addition, we have computed the transmission for a thin planar film with triangular holes (with the same size as those in the experiment) in graphene symmetry. Again transmission values are low when compared with those in the capped sphere system. These findings suggest

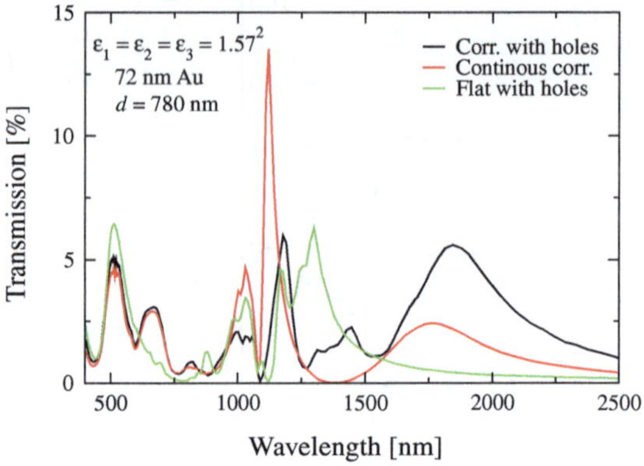

Fig. 2.24 Calculated transmission curves for corrugated (with and without holes) and planar metal film with holes in a graphene symmetry. All films were modeled in a homogeneous media. The metal considered is Au and the lattice parameter is d = 780 nm. From Ref. [16].

that the presence of the photonic crystal layer is of great importance in the overall transmission mechanism. The close spectral correspondence between transmission peaks in the coated case and transmission dips in the uncoated one, already points to the possible relevance of guided modes in the photonic crystal. This relevance is corroborated by the computed electromagnetic field distributions (see Fig. 2.25 for a representative case), which present strong field confinement at the location of the spheres.

It is interesting to highlight the differences on the transmittance between guided modes in a photonic crystal and guided modes in a uniform dielectric slab. The first difference is related to the "energetics". A first estimation of the spectral position at which EOT features appear can be obtained by computing the frequency of the surface mode involved, at a wavevector equal to the shortest reciprocal lattice vector (for the case of normal incidence considered here). Similarly, dips in the corrugated dielectric are expected to appear at the same condition, as Fig. 2.21 shows. Let us start by considering the uniform dielectric slab. The point here is that the guided modes in a vacuum-metal-dielectric film-substrate (VMDS) waveguide are different from the ones in a vacuum-dielectric film-substrate (VDS) configuration, due to the large differences between the Fresnel coefficients for metal/dielectric and metal/air interfaces. Therefore, features in a corrugated dielectric and a corrugated metal, each of them placed on top of dielectric slab, should appear at different wavelengths. To illustrate this point, we have computed the wavelengths of the guided modes in both VMDS and VDS configurations, for the following parameters (motivated by the experimental setup): the dielectric film has a dielectric constant $\varepsilon = 1.57^2$ and a thickness $t = 780$ nm. The substrate has a dielectric constant $\varepsilon = 1.5^2$. The considered wavevector is $k = 2\pi/t$ (in a sphere array the inter-distance between spheres is equal to the dielectric film thickness). The metal thickness is 70 nm and its dielectric constant is taken as $\varepsilon_{\mathrm{metal}} = -50$ (approximately the value for Au at $\lambda \sim 1000$ nm). We obtain that the wavelengths of the guided modes are: 1176 nm for the VDS configuration and 1235 nm for the VMDS case. On the contrary, the dispersion relation of guided modes in the photonic crystal (composed by the two-

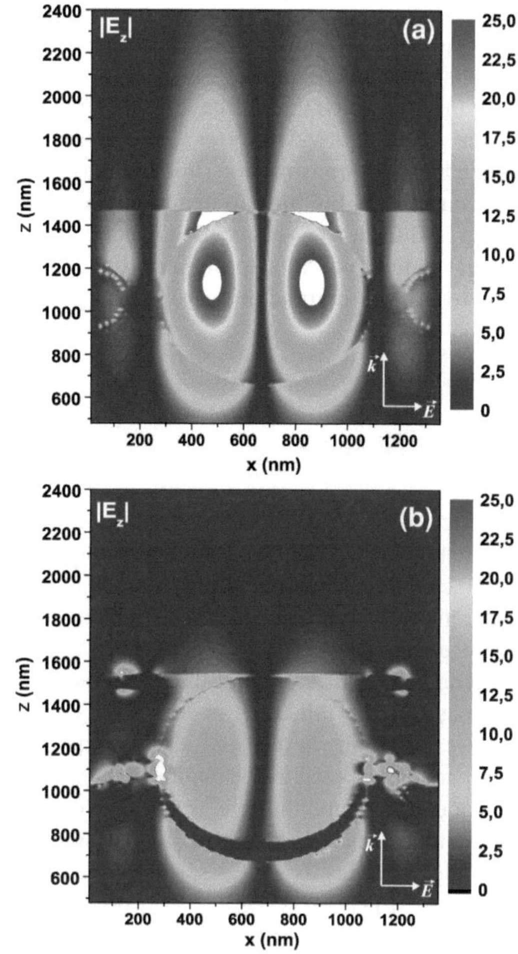

Fig. 2.25 Contour plots for the modulus of the z-component of the electric field across a plane passing through the center of the spheres. The system under study is the one considered in 2.21. (**b**). **a** uncoated case at the wavelength of the main transmission dip ($\lambda = 944$ nm). **b** coated case at the main transmission maximum ($\lambda = 1002$ nm). From Ref. [16].

dimensional arrays of dielectric spheres) is weakly affected by the presence of the metal film (calculations not presented here estimate that the difference between the wavelengths of the guided mode in the metal capped and uncapped configurations is of the order of 5 nm). This is so because, in this case, the z-component of the electric field (which is the relevant one for guided modes) is more concentrated close to the center of the spheres (see Fig. 2.25), so a smaller fraction of the field senses the different Fresnel coefficients alluded above.

The second difference is related to the coupling of the light, passing through the metal film in the presence of guided modes, to the different radiation orders. Guided modes in photonic crystals represent a weaker coupling to radiation modes than either guided modes in a dielectric or surface plasmons (again due to the previously cited concentration of the electric field in the photonic crystal guided modes, which places the EM field away from the radiation region). Notice that radiation damping impairs the resonant transmission process, so this feature of photonic crystal modes explains why the configuration of metal film on top of a photonic crystal is so efficient for EOT phenomena (Compare Figs. 2.23 with 2.24).

It must be noted that we have concentrated on the transmission peak appearing at larger wavelengths. At shorter (optical) wavelengths there is also a close correspondence between transmission dips in the uncoated system and transmission peaks in the coated one, pointing to again to the relevance of guided modes. These modes could be due either to remappings (aided by a reciprocal lattice vector) of the fundamental guided mode or to higher order guided modes. No attempt has been made here to assign a definite origin to these modes as they give rise to small transmission peaks.

2.6 Conclusions

In conclusion, we have investigated three different questions that strongly affect EOT: the metal chosen, the metal thickness and the hole shape.

To elucidate about the influence of the metal chosen on EOT we have investigated theoretically the resonant optical transmission through circular hole arrays drilled in different metals. We have performed two series of calculations. In both of them all lengths except the metal thickness are scaled by the same factor. In one of the series the metal thickness is kept fixed, while in the other the metal thickness is also scaled. In the first case, for which there is experimental data available [32], the comparison between experiment and theory is very good. These results confirm that, in the experiments, the effects of possible inhomogeneities in hole shape and size were small. Also the metal surface, despite the processing that has received when creating the holes, is well described by a dielectric constant close to its bulk value. Moreover, FDTD is validated as a predictive tool for this kind of systems, as the comparison with experimental data did not require any fitting parameter. The case in which all lengths were scaled allowed the comparison of the transmission properties of real metals with those of a perfect conductor. The analysis reveals different types of behavior of the transmittance in hole arrays in different sets of metals: in Ag, Au or Cu, the transmittance is even larger than in the perfect conductor case, reflecting that absorption is low and the penetration of EM fields effectively enlarges the hole area. In Ni and Cr, although the effective area is as large as in the previous metals, absorption strongly reduces the resonant transmittance. Aluminum behaves very much like a perfect conductor at the lower wavelength end of the optical regime but for peak wavelengths $\geq$700 nm the resonant transmittance characteristics are dominated by absorption, as in the case of Ni and Cr. Finally, tungsten, which in the spectral range considered is a dielectric, presents transmission resonances with maxima much smaller than those of even the worse metals (Ni, Cr).

In the study of EOT when the film thickness is varied, we have shown that the EOT peak can be tuned to longer wavelengths (by even hundreds of nm) by decreasing the film thickness without strongly affecting neither transmission intensity nor peak visibility (which is still large at $w \sim 20$ nm). We have demonstrated that only SRs modes are responsible for the EOT phenomenon in optically thin metallic 2DHAs. This may be of interest in the fields of EOT and Negative Refractive Index (which has been obtained in stacked optically thin 2DHAs [82]).

From our study on the hole shape dependence of EOT, we have explained theoretically the interplay between two different mechanisms that enhance the transmission of light through 2D arrays of rectangular holes: SPP-based extraordinary transmission and enhanced transmission assisted by the excitation of a localized resonance, spectrally located at the cut-off wavelength of the hole waveguide, λ_c. We have shown that when $d < \lambda_c$ the transmission resonance has a localized nature mainly, i.e. it is mainly governed by the behavior of a single hole. In contrast, for $d > \lambda_c$, SPP governs the transmittance through the structure, We have also demonstrated that in this last case, resonant transmission is mainly limited by the absorption in the metal.

Finally, the 2D-photonic crystal structure composed by periodically arranged microspheres and covered with thin metal films, has been found to present EOT. Measured spectra have been compared with spectra calculated with FDTD and the good agreement has allowed modeling of slightly modified structures to get further information about possible transmission mechanisms. The calculations indicate that the guided modes in the PCS are mainly responsible to the relatively large transmission values observed (especially for the main peak). In contrast, the small holes in the thin metal film (at the interstices between three adjacent spheres) and metal deposit onto the support do not strongly influence the main transmission peak. The high transmission values, straightforward fabrication and easy up-scaling of the metal covered slabs together with simple peak positioning in a broad wavelength region (VIS/IR) make these structures a good candidate for application purposes.

References

1. H.A. Bethe, Phys. Rev. **66**, 163 (1944)
2. T.W. Ebbesen, H.L. Lezec, H.F. Ghaemi, T. Thio, P.A. Wolff, Nature **391**, 667 (1998)
3. H.F. Ghaemi, T. Thio, D.E. Grupp, T.W. Ebbesen, H.J. Lezec, Phys. Rev. B **58**, 6779 (1998)
4. L. Martín-Moreno, F.J. García-Vidal, H.J. Lezec, K.M. Pellerin, T. Thio, J.B. Pendry, T.W. Ebbesen, Phys. Rev. Lett. **86**, 1114 (2001)
5. L. Martín-Moreno, F.J. García-Vidal, Opt. Express **12**, 3619 (2004)
6. C. Genet, T.W. Ebbesen, Nature **445**, 39 (2007)
7. D.E. Grupp, H.J. Lezec, T.W. Ebbesen, K.M. Pellerin, T. Thio, Appl. Phys. Lett. **77**, 1569 (2000)
8. Q. Wang, J. Li, C. Huang, C. Zhang, Y. Zhu, Appl. Phys. Lett. **87**, 091105 (2005)
9. J.R. Krenn, A. Dereux, J.C. Weeber, E. Bourillot, Y. Lacroute, J.P. Goudonnet, G. Schider, W. Gotschy, A. Leitner, F.R. Aussenegg et al., Phys. Rev. Lett. **82**, 2590 (1999)
10. H.J. Lezec, A. Degiron, E. Devaux, R.A. Linke, L. Martín-Moreno, F.J. García-Vidal, T.W. Ebbesen, Science **297**, 820 (2002)
11. de F.J.G. Abajo, Rev. Mod. Phys. **79**, 1267 (2007)
12. F. J. Garcia-Vidal, L. Martin-Moreno, T. W. Ebbesen, and L. K. Kuipers, Rev. Mod. Phys. (2009), (To be published).
13. S.G. Rodrigo, F. J. García-Vidal, L. Martín-Moreno, Phys. Rev. B **77**, 075401 (2008)

14. S.G. Rodrigo, L. Martín-Moreno, A.Y. Nikitin, A.V. Kats, I.S. Spevak, F.J. García-Vidal, Opt. Lett. **34**, 4 (2009)
15. A. Mary, S.G. Rodrigo, L. Martín-Moreno, F.J. García-Vidal, Phys. Rev. B **76**, 195414 (2007)
16. L. Landström, D. Brodoceanu, D. Bäuerle, F.J. Garcia-Vidal, S.G. Rodrigo, L. Martin-Moreno, Opt. Express **17**, 761 (2009)
17. L. Salomon, F.D. Grillot, A.V. Zayats, de F. Fornel, Phys. Rev. Lett. **86**, 1110 (2001)
18. P. Lalanne, J.C. Rodier, J.P. Hugonin, J. Opt. A: Pure Appl. Opt. **7**, 422 (2005)
19. C. Genet, van M.P. Exter, J.P. Woerdman, Opt. Commun. **225**, 331 (2003)
20. M. Sarrazin, J.-P. Vigneron, Phys. Rev. E **68**, 016603 (2003)
21. E. Moreno, L. Martín-Moreno, F.J. García-Vidal, J. Opt. A: Pure Appl. Opt. **8**, S94 (2006)
22. E. Moreno, F.J. García-Vidal, L. Martín-Moreno, Phys. Rev. B. **69**, 121402(R) (2004)
23. B. Wang, W. Dai, A. Fang, L. Zhang, G. Tuttle, T. Koschny, C.M. Soukoulis, Phys. Rev. B **74**, 195104 (2006)
24. E. Dulkeith, S.J. McNab, Y.A. Vlasov, Phys. Rev. B **72**, 115102 (2005)
25. J. Christensen, A.I. Fernández-Domínguez, de F. León-Pérez, L. Martín- Moreno, F.J. García-Vidal, Nature Phys. **3**, 851 (2007)
26. E. Moreno, A.I. Fernández-Domínguez, J.I. Cirac, F.J. García-Vidal, L. Martín-Moreno, Phys. Rev. Lett. **95**, 170406 (2005)
27. R. Gordon, A.G. Brolo, A. McKinnon, A. Rajora, B. Leathem, K.L. Kavanagh, Phys. Rev. Lett. **92**, 037401 (2004)
28. K.J.K. Koerkamp, S. Enoch, F.B. Segerink, van N.F. Hulst, L. Kuipers, Phys. Rev. Lett. **92**, 183901 (2004)
29. A. Degiron, T.W. Ebbesen, J. Opt. A: Pure Appl. Opt. **7**, S90 (2005)
30. A. Degiron, H.J. Lezec, N. Yamamoto, T.W. Ebbesen, Opt. Commun. **239**, 61 (2004)
31. F.J. García-Vidal, E. Moreno, J.A. Porto, L. Martín-Moreno, Phys. Rev. Lett. **95**, 103901 (2005)
32. F. Przybilla, A. Degiron, J.-Y. Laluet, C. Genet, T.W. Ebbesen, J. Opt. A: Pure Appl. Opt. **8**, 458 (2006)
33. A. Vial, A.-S. Grimault, D. Macías, D. Barchiesi, de la M.L. Chapelle, Phys. Rev. B **71**, 085416 (2005)
34. A.D. Rakić, A.B. Djurišić, J.M. Elazar, M.L. Majewski, Appl. Opt. **37**, 22 (1998)
35. E.D. Palik, *Handbook of optical constants of solids* ed. by Palik, D. Edward (Academic Press, New York, 1985)
36. E.D. Palik, *Handbook of optical constants of solids II* ed. by Palik, D. Edward (Academic Press, Boston, 1991).
37. A. Krishnan, T. Thio, T.J. Kim, H.L. Lezec, T.W. Ebbesen, P.A. Wolff, J.B. Pendry, L. Martín-Moreno, F.J. García-Vidal, Opt. Commun. **200**, 1 (2001)
38. J.B. Pendry, L. Martín-Moreno, F.J. García-Vidal, Science **305**(5685), 847 (2004)
39. F.J. García-Vidal, L. Martín-Moreno, J.B. Pendry, J. Opt. A: Pure Appl. Opt. **7**, S97 (2005)
40. de F.J.G. Abajo, J.J. Saenz, Phys. Rev. Lett. **95**, 233901 (2005)
41. J. Bravo-Abad, F.J. García-Vidal, L. Martín-Moreno, Phys. Rev. Lett. **93**, 227401 (2004)
42. F.J. García-Vidal, L. Martín-Moreno, E. Moreno, .K.S. Kumar, R. Gordon, Phys. Rev. B **74**, 153411 (2006)
43. M. Sarrazin, J.-P. Vigneron, Phys. Rev. B **71**, 075404 (2005)
44. I.R. Hooper, J.R. Sambles, Phys. Rev. B **70**, 045421 (2004)
45. D. Gérard, L. Salomon, de F. Fornel, A.V. Zayats, Opt. Express **12**, 3652 (2004)
46. E.N. Economou, Phys. Rev. **182**, 539 (1969)
47. A. Azad, W. Zhang, Opt. Lett. **30**, 2945 (2005)
48. X. Shou, A. Agrawal, A. Nahata, Opt. Express **13**, 9834 (2005)
49. Q. Cao, P. Lalanne, Phys. Rev. Lett. **88**, 057403 (2002)
50. S.A. Darmanyan, A.V. Zayats, Phys. Rev. B **67**, 035424 (2003)
51. A.V. Kats, M.L. Nesterov, A.Y. Nikitin, Phys. Rev. B **76**, 045413 (2007)
52. H. Cao, A. Nahata, Opt. Express **12**, 3664 (2004)

53. van der K.L. Molen, K.J.K. Koerkamp, S. Enoch, F.B. Segerink, van N.F. Hulst, L. Kuipers, Phys. Rev. B **72**, 045421 (2005)
54. Z. Ruan, M. Qiu, Phys. Rev. Lett. **96**, 233901 (2006)
55. R. Gordon, A. Brolo, Opt. Express **13**, 1933 (2005)
56. W. Jia, X. Liu, Eur. Phys. J. B **46**, 343 (2005)
57. T. López-Rios, D. Mendoza, F.J. García-Vidal, J. Sánchez-Dehesa, B. Pannetier, Phys. Rev. Lett. **81**, 665 (1998)
58. D. Crouse, M. Arend, J. Zou, P. Keshavareddy, Opt. Express **14**, 2047 (2006)
59. D.C. Skigin, R.A. Depine, Phys. Rev. E **63**, 046608 (2001)
60. de F.J.G. Abajo, R. Gómez-Medina, J.J. Sáenz, Phys. Rev. E **72**, 016608 (2005)
61. D.E. Grupp, H. Lezec, T. Thio, T. Ebbesen, Adv. Mater. **11**, 860 (1999)
62. Y.-H. Ye, J.-Y. Zhang, Appl. Phys. Lett. **84**, 2977 (2004)
63. van der K.L. Molen, F.B. Segerink, van N.F. Hulst, L. Kuipers, Appl. Phys. Lett. **85**, 4316 (2004)
64. D. Qu, D. Grischkowsky, Phys. Rev. Lett. **93**, 196804 (2004)
65. C. Janke, J.G. Rivas, C. Schotsch, L. Beckmann, P.H. Bolivar, H. Kurz, Phys. Rev. B **69**, 205314 (2004)
66. W.L. Barnes, W.A. Murray, J. Dintinger, E. Devaux, T. Ebbesen, Phys. Rev. Lett. **92**, 107401 (2004)
67. L. Landström, D. Brodoceanu, N. Arnold, K. Piglmayer, D. Bäuerle, Appl. Phys. A **81**, 911 (2005)
68. L. Landström, N. Arnold, D. Brodoceanu, K. Piglmayer, D. Bäuerle, Appl. Phys. A **83**, 271 (2006)
69. L. Landström, D. Brodoceanu, K. Piglmayer, D. Bäuerle, Appl. Phys. A **84**, 373 (2006)
70. L. Collot, V. Lefevre-Seguin, M. Brune, J. Raimond, S. Haroche, Europhys. Lett. **23**, 327 (1993)
71. M.-H. Wu, G.M. Whitesides, Appl. Phys. Lett. **78**, 2273 (2001)
72. D. Bäuerle, T. Gumpenberger, D. Brodoceanu, G. Langer, J. Kofler, J. Heitz, K. Piglmayer, in *Laser Cleaning II*, ed. by D.M. Kane (World Scientific, Singapore, 2005).
73. D. Bäuerle, G. Wysocki, L. Landström, J. Klimstein, K. Piglmayer, J. Heitz, Proc. SPIE **5063**, 8 (2003)
74. K. Piglmayer, R. Denk, D. Bäuerle, Appl. Phys. Lett. **80**, 4693 (2002)
75. G. Wysocki, R. Denk, K. Piglmayer, N. Arnold, D. Bäuerle, Appl. Phys. Lett. **82**, 92 (2003)
76. D. Bäuerle, K. Piglmayer, R. Denk, N. Arnold, Lambda Highlights **60**, 1 (2002)
77. D. Bäuerle, *Laser Processing and Chemistry*, 3rd edn. (Springer, Berlin, 2000)
78. L. Landström, J. Klimstein, G. Schrems, K. Piglmayer, D. Bäuerle, Appl. Phys. A **78**, 537 (2004)
79. G. Langer, D. Brodoceanu, D. Bäuerle, Appl. Phys. Lett. **89**, 261104 (2006)
80. A. Pikulin, N. Bityurin, D. Brodoceanu, G. Langer, D. Bäuerle, Appl. Phys. Lett. **91**, 191106 (2007)
81. R. Micheletto, H. Fukuda, M. Ohtsu, Langmuir **11**, 3333 (1995)
82. S. Zhang, W. Fan, N.C. Panoiu, R.M. Osgood, S.R.J. Brueck, Phys. Rev. Lett. **95**, 137404 (2005)

Chapter 3
Theory of Negative-Refractive-Index Response of Double-Fishnet Structures

3.1 Introduction

In the last years we have seen how a new sort of man-made materials with "exotic" optical properties (not found in nature) could be designed. In turn, this has opened amazing prospects for future technological applications. One relates to the design of structures with negative refractive index (NRI). In these NRI metamaterials, firstly proposed by Veselago [1], both the electric permittivity and the magnetic permeability are negative, leading to a refractive index that is negative. Metamaterials presenting NRI are expected to lead to important applications, as the perfect lens proposed by Pendry [2]. Furthermore, NRI metamaterials are thought to be the bricks for solving a long lasting question in optics: Is it possible to make an object "invisible"? Predicted independly by Ulf Leonhardt [3] and Pendry et al. [4], such an "invisibility cloak" was demonstrated in the microwave regime [5].

Veselago demonstrated that an isotropic, homogeneous and lineal (i.h.l) medium characterized by negative values of both the permittivity (ε) and the permeability (μ), does not contradict any fundamental law of physics. In fact, a substance like that behaves in a completely different fashion that a conventional material (See Ref. [6] for a complete review).

Plane waves traveling through i.h.l media at fixed frequency (ω) can be expressed as follows: $\vec{E}(\vec{r},t) = \vec{E}_0 e^{\imath(\vec{k}\vec{r}-\omega t)}$ and $\vec{H}(\vec{r},t) = \vec{H}_0 e^{\imath(\vec{k}\vec{r}-\omega t)}$. In this case, by applying Maxwell's equations we obtain:

$$\begin{aligned} \vec{k} \times \vec{E}_0 &= \mu\mu_0\omega\vec{H}_0 \\ \vec{k} \times \vec{H}_0 &= -\varepsilon\varepsilon_0\omega\vec{E}_0 \end{aligned} \tag{3.1}$$

in the international system of units (MKS). As usual, $\omega^2 = \frac{c^2|k|^2}{\varepsilon\mu}$, being c the speed of light in vacuum. From Eq. 3.1 obvious that the set $\{\vec{k}, \vec{E}, \vec{H}\}$ defines a right-handed (RH) triplet of vectors whether $\varepsilon > 0$ and $\mu > 0$. Conversely, if $\varepsilon < 0$ and $\mu < 0$ they constitute a left-handed (LH) set. Note that Eq. 3.1 remain invariant by changing

S. G. Rodrigo, *Optical Properties of Nanostructured Metallic Systems*,
Springer Theses, DOI: 10.1007/978-3-642-23085-1_3,

the signs of the permittivity, the permeability and the $\vec{k}$-vector ($\vec{k} \rightarrow -\vec{k}$, which defines the phase velocity), simultaneously. Hence, a plane wave will propagate within LH material with opposite phase velocity as it were traveling through a RH one. The energy flux carried by a plane wave is determined by the Poynting vector $\vec{S}$, which is given by

$$\vec{S} = \vec{E} \times \vec{H} \tag{3.2}$$

The $\vec{S}$ vector always forms a RH set with $\vec{E}$ and $\vec{H}$, thus the flow of energy is opposite to the phase velocity in LH media. We arrive at the same conclusion taking into account that $\vec{S} = \frac{|\vec{E}|^2}{\mu\mu_0\omega}\vec{k} = \frac{|\vec{H}|^2}{\epsilon\epsilon_0\omega}\vec{k}$.

Another important discussion is about the energy carried by an EM field in LH materials. The total energy of an EM field at point in space reads, $U = \frac{1}{2}\varepsilon\vec{E}^2 + \frac{1}{2}\mu\vec{H}^2$, for i.h.l and non-dispersive media. Clearly, this expression is not compatible with negative values of ε and μ. Therefore, both quantities must depend on frequency, and the proper definition of total energy is given by $U = \frac{1}{2}\frac{\partial(\varepsilon\omega)}{\partial\omega}\vec{E}^2 + \frac{1}{2}\frac{\partial(\mu\omega)}{\partial\omega}\vec{H}^2$.

Up to here, we have revisited the propagation properties of light inside LH materials. Note that if the whole universe were made of LH materials redefining the phase velocity ($\vec{k} \rightarrow -\vec{k}$), all the electromagnetic phenomena would be exactly the same as in a RH universe. In the end, it would be a matter of agreement as regards to the sign of the parameters (ε, μ, $\vec{k}$, ...). Really, the key point is how light behaves passing from a RH medium to a LH medium, or viceversa. When light travels from a medium to another, the boundary conditions for EM fields,

$$\begin{aligned} E_t^1 &= E_t^2 \\ H_t^1 &= H_t^2 \\ \varepsilon_1 E_n^1 &= \varepsilon_2 E_n^2 \\ \mu_1 H_n^1 &= \mu_2 H_n^2 \end{aligned} \tag{3.3}$$

must be satisfied, independently of whether or not the media have the same "rightness" (integer numbers label different media). The superscripts t and n refers to the tangential and normal to the surface field components, respectively. In the RH medium the incoming ray will be labeled with a i subscript and r for the reflected one. As we can deduce from the boundary conditions [Eq. 3.3], the tangential field components maintain their directions, whereas the normal components change, and not only their relative values, but their directions as well. We depict in Fig. 3.1 the refraction-reflection construction for a TM-plane wave traveling from a RH medium to a LH medium. As we can see, the Poynting vector (represented by dashed arrows) flows from the RH medium to the LH medium so that energy is conserved. Taking into account both the proper direction the $\vec{k}$-vector follows in each media and the field boundary conditions, only the ray refracted to the left of Fig. 3.1 has physical meaning.

Additionally, it can be seen from Fig. 3.1 that the usual Snell's law should be given more precisely,

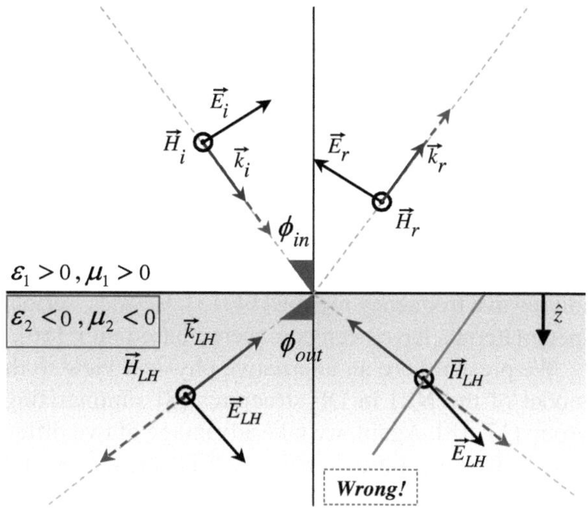

Fig. 3.1 Two possible paths to follow a ray passing through the boundary between a RH medium and a LH one. Only the refracted ray rendered to the *left* has physical sense

$$RH \rightarrow RH \tag{3.4}$$

$$\frac{sin(\phi_{in})}{sin(\phi_{out})} = \sqrt{\frac{\varepsilon_2\mu_2}{\varepsilon_1\mu_1}} \tag{3.5}$$

$$RH \rightarrow LH \tag{3.6}$$

$$\frac{sin(\phi_{in})}{sin(\phi_{out})} = -\sqrt{\frac{\varepsilon_2\mu_2}{\varepsilon_1\mu_1}} \tag{3.7}$$

showing the difference between RH refraction ($\phi_{out} > 0$) and LH refraction ($\phi_{out} < 0$). In particular, the index of refraction of a LH medium relative to vacuum is negative, so they are called NRI media. That negative values of ε and μ give raise to negative values of n can be also demonstrated as follows. Let us write $\varepsilon = \varepsilon_r + \imath\varepsilon_i$ and $\mu = \mu_r + \imath\mu_i$, where we take into account the absorbing properties of the medium ($\varepsilon_i > 0$ and $\mu_i > 0$). The refractive index is $n = \pm\sqrt{\varepsilon\mu}$ or explicitly, $n = \pm\sqrt{(\varepsilon_r + \imath\varepsilon_i)(\mu_r + \imath\mu_i)}$. This expression can be approximated $n \approx \pm\sqrt{\varepsilon_r\mu_r + \imath(\varepsilon_i\mu_r + \mu_i\varepsilon_r)} \approx \pm\sqrt{\varepsilon_r\mu_r}[1 + \imath\frac{\varepsilon_i\mu_r + \varepsilon_r\mu_i}{\varepsilon_r\mu_r}]$. The radiation condition (the field must vanish at infinity) imposes that $n_i > 0$, so if $\varepsilon_r < 0$ and $\mu_r < 0$ at the same time,then the negative sign must be chosen.

No natural material known possesses negative magnetic permeability. To date, the only way to achieve the above-mentioned behavior is by combining different materials, built with different geometries. The system arising is usually called a "meta-material" because its optical response may be different than the optical response of its components in bulk. An example of a meta-material displaying negative μ was first reported by Pendry et al. [7]. They proposed to build arrays of split-ring resonators. The effective negative magnetic susceptibility found in them paved the way in the search of structures presenting NRI (For a recent review on negative-index metamaterials see Ref. [8]).

In 2005 Zhang et al. [9, 10] proposed and demonstrated a negative index metamaterial working at near infrared frequencies with a design very similar to the structure showing EOT. This metamaterial is composed by a two-dimensional (2D) array of holes penetrating completely in a metal-dielectric-metal film stack, the so-called double-fishnet (DF) structure. This structure has received a lot of attention for its optical response as NRI material at visible [11] and near infrared frequencies [12, 13]. Other studies have shown that this NRI metamaterial design also operates in the microwave frequency regime [14, 15]. Besides, serious attempts to get truly 3D-NRI metamaterials have been engineered based on it [16].

We present here an alternative physical view to the well established "circuitry" model of the NRI in DF structures [8] summarizing two works published in our group [17, 18]. Again, we take advantage of two different theoretical techniques: the finite-difference time-domain (FDTD) method Sect. 1.2 and an adapted version of the coupled mode method (CMM) Sect. 1.3 for the DF structure, which was carried out by A. Mary at the Departamento de Física Teórica de la Materia Condensada (Universidad Autónoma de Madrid). Within these formulation, these structures can be interpreted as holey plasmonic metamaterials instead of wire fabrics.

3.2 Theory of Negative-Refractive-Index Response of Double Fishnet Structures

A DF structure can be described as three films (metal-dielectric-metal) drilled with a square array of rectangular holes, with h_m and h_d being the thickness of the metallic films and the dielectric layer, respectively. In our study, we analyze two very distinct frequency regimes. First, we consider that the metal behaves as a perfect electrical conductor (PEC), which is a very good approximation for metals at microwave or terahertz frequencies. When analyzing this limit, we will use the period, d, as the unit length. We also present results for silver at optical and near infrared frequencies. In this case, we take the dielectric function of silver from Table 1.1 and we choose $d = 600\,\text{nm}$. The structures are illuminated by a p-polarized plane wave (i.e., the in-plane component of the incident E-field pointing along the short edge of the holes). When analyzing perforated silver films at optical frequencies, we use the finite difference time domain (FDTD) method (See Sect. 1.2) which, as previously described, is virtually exact provided that a small enough mesh-size is chosen to account for the rapid variations of the EM fields. Along with the FDTD method we also use CMM within the surface impedance approximation (SIBC) (See Sect. 1.3). In order to analyze 2DHA and DF structures made of PEC material, we apply CMM as well. The great advantage of CMM is that, when dealing with subwavelength holes, a very good approximation to the transmission and reflection properties can be achieved by considering only the least decaying evanescent mode inside the holes (the TE_{01} mode for the chosen polarization of the incident plane wave). This allows a semi-analytical treatment and to extend the formalism to study a large number of metallic films, as done at the end of this chapter, that would be difficult with the FDTD method. For a complete description of CMM applied to DF structures see Ref.[18].

3.2.1 *Effective Parameters of 2DHAs*

Schematic pictures of the two systems under study (2DHA and DF) are shown in the insets of figs 3.2a and 3.3a, respectively. The 2DHA structure is an infinite square array (period d) of rectangular holes of sides a_x and a_y, perforated on a freestanding metallic film of thickness h_1. To further simplify the analysis without loosing generality, the dielectric between the two metal films is chosen to be vacuum or air. The geometrical parameters are: $a_x = 0.33\ d$, $h_m = 0.05\ d$, $h_d = 0.05\ d$ and $h_1 = 2h_m + h_d = 0.15\ d$. The long edge of holes a_y will vary between 0.33d and 0.98d. These values are representative of those used in the experimental samples showing NRI [12].

First, we revisit the transmission properties of a 2DHA made of rectangular holes. Figure 3.2a renders the corresponding normal incidence transmittance spectra as a function of wavelength (λ) and a_y for the PEC case. For small holes, two transmission peaks (leading to 100 % transmission), whose spectral locations appear close to d, emerge in the spectrum. These two resonances correspond to the symmetric and antisymmetric combinations of the two surface EM modes associated with the two interfaces of the structure [19]. Note that, as the metal behaves as a PEC, these modes are not real SPPs but geometry-induced surface EM modes, the so-called *spoof* SPPs [20, 21]. As a_y is increased, the extremely narrow peak associated with the antisymmetric combination remains unaltered at $\lambda \approx d$. However, the symmetric combination of SPPs strongly hybridizes with the cutoff resonance, which has been discussed in Sect. 2.4. For rectangular holes appears close to the cut-off wavelength of the fundamental mode (TE_{01} mode) of the hole waveguide, $\lambda_C = 2a_y$. From now on, we name *cutoff resonance* to this hybridized mode that is combination of the symmetric surface EM mode and the Fabry-Perot resonance. From the knowledge of the zero-order transmission and reflection coefficients, it is possible to retrieve bi-univocally the effective optical parameters [22]. The method described in Ref. [22] is reliable provided the zero-order diffracted beams are the only ones propagating in both the reflection and transmission regions (for normal incidence this implies $\lambda > d$). The real part of the effective permittivity, ϵ_{eff}, of the 2DHA structure is shown in Fig. 3.2b. Dashed and solid lines indicate the wavelength in which $Re[\epsilon_{\text{eff}}] = 0$ and the cutoff wavelength, respectively. Note that except in the region $\lambda \approx d$, there is a close correspondence between these two wavelengths. This kind of coincidence can be understood by revisiting the concept of spoof SPP modes. As explained in Ref. [20], the dielectric response (in the effective medium limit in which diffraction effects are neglected) of a semi-infinite holey PEC presents a Drude-like behavior in

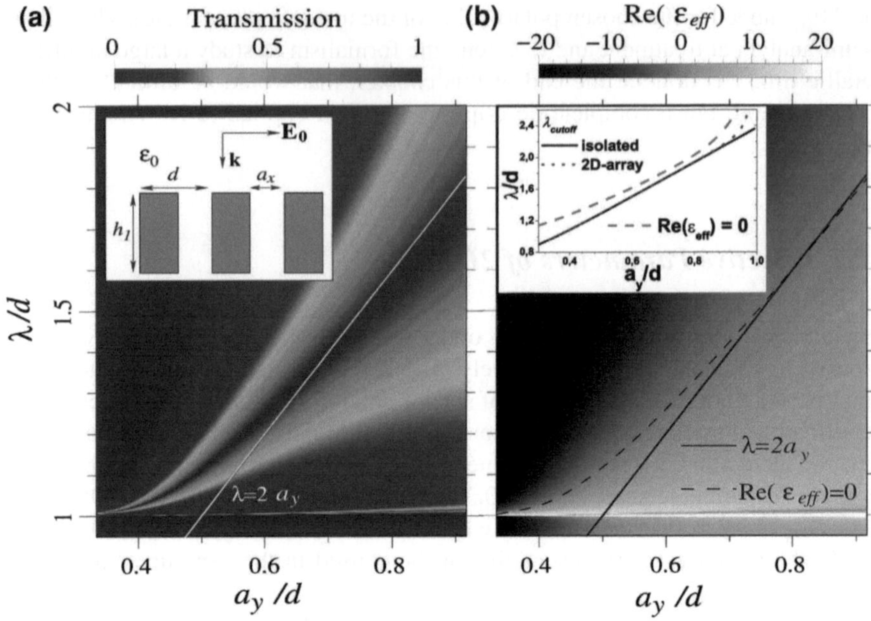

Fig. 3.2 **a** Transmittance versus both a_y and λ for a 2DHA perforated on a PEC film, with geometrical parameters: $a_x = 0.33\ d$ and $h_1 = 0.15\ d$. Inset contains an schematic picture of the 2DHA. **b** Re $[\epsilon_{\text{eff}}]$ as a function of λ and a_y. *Solid line* Cutoff wavelength, $\lambda = 2a_y$. *Dashed line* λ that satisfies the condition Re $[\epsilon_{\text{eff}}(\lambda)] = 0$. In the inset, the corresponding FDTD results for silver-2DHA with the same parameters as in **a** and $d = 600$ nm are displayed.

which the plasma frequency of the electron gas is replaced by the cut-off frequency of the hole waveguide. Regarding the magnetic response, the effective magnetic permeability, μ_{eff} takes a constant value. In mathematical terms, these two last sentences translate into:

$$\epsilon_{\text{eff}} = \frac{1}{S^2}\left(1 - \frac{\omega_p^2}{\omega^2}\right), \quad \mu_{\text{eff}} = S^2 \tag{3.8}$$

where $S = 2\sqrt{2\,a_x\,a_y}/\pi d$ and $\omega_p = \pi c/a_y$. Therefore, in the effective medium approximation ($\lambda >> d$), a holey *semi-infinite* PEC is characterized by a $Re[\epsilon_{\text{eff}}]$ that changes from negative ($\lambda < \lambda_C$) to positive values ($\lambda > \lambda_C$), and by a μ_{eff} that remains constant and positive (Here $\lambda_C = 2a_y$).

Figure 3.2b shows that the relation between λ_C and the condition $Re[\epsilon_{\text{eff}}] = 0$ also holds for a very thin PEC film and for λ slightly larger than d. Moreover, this link is maintained when moving to frequencies in the optical regime, as shown in the inset of Fig. 3.2b. Here the spectral locations of both the cutoff wavelength of the hole waveguide (black line) and the condition $Re[\epsilon_{\text{eff}}] = 0$ (green dashed line) are rendered as a function of a_y, for the same geometrical parameters as in panel (a),

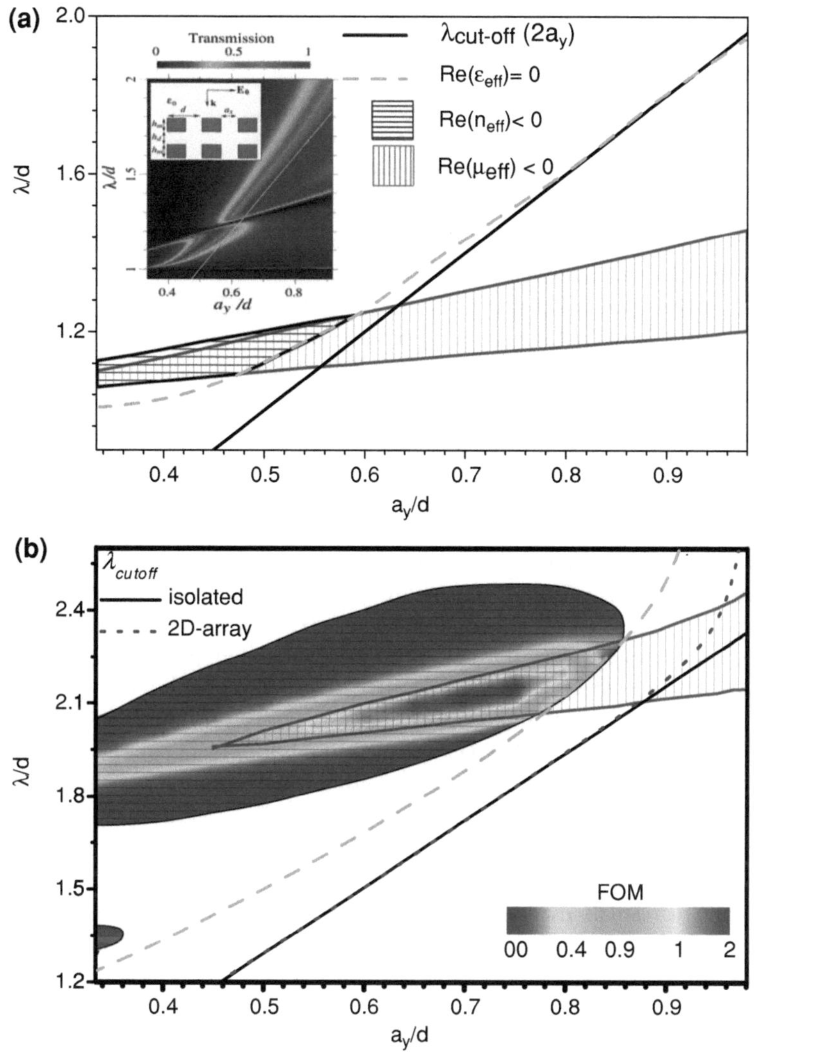

Fig. 3.3 **a** The evolution of the spectral locations of both the cutoff wavelength (*black line*) and the condition $Re[\epsilon_{\text{eff}}] = 0$ (*green dashed line*). The regions of negative $Re[\mu_{\text{eff}}]$ (*red vertical lines*) and $Re[n_{\text{eff}}]$ (*black horizontal lines*) are also displayed as a function of a_y for the PEC-DF case. Inset: transmittance versus both a_y and λ for a PEC-DF structure with the geometrical parameters as in Fig. 3.2. *Solid curve* renders $\lambda = 2a_y$ line. **b** As in **a**, but for a silver-DF structure with $d = 600$ nm. A contour plot of the FOM is rendered, for the regions where $Re[n_{\text{eff}}] < 0$

and $d = 600$ nm. Notice however that for a silver-2DHA in the limit $a_y \to d$, the condition $Re[\epsilon_{\text{eff}}] = 0$ is linked to the cutoff wavelength calculated with the FDTD method for waveguides forming a 2D periodic array (blue dashed line in the inset of

Fig. 3.2b). In this limit, the cutoff deviates from the one of an isolated waveguide, due to the cross-talk between waveguides through the vertical metal walls.

3.2.2 The Double-Fishnet Structure

The rest of this chapter is devoted to analyze the optical properties of the DF structure. Inset of Fig. 3.3a renders the normal incidence transmission spectra for the PEC case. The two transmission peaks previously described for a 2DHA also appear in the DF structure. Fig. 3.3a shows the spectral location of the condition $Re[\epsilon_{\text{eff}}] = 0$ (green dashed line), and the regions in which $Re[\mu_{\text{eff}}]$ (red vertical lines) and $Re[n_{\text{eff}}]$ (black horizontal lines) are negative. Notice that the cutoff resonance wavelength also controls $Re[\epsilon_{\text{eff}}]$ in the case of a PEC-DF structure, as shown in Fig. 3.3a. On the other hand, the presence of a second metal layer and a dielectric gap results in the emergence of additional resonant features (transmission peak and dip) in the transmission spectra. Associated with this new resonant behavior, $Re[\mu_{\text{eff}}]$ presents negative values (see panel (a)). As expected, the effective refractive index is negative when both $Re[\epsilon_{\text{eff}}]$ and $Re[\mu_{\text{eff}}]$ are negative. This condition is only satisfied when $a_y < 0.58\ d$ (for this set of geometrical parameters). For larger holes, as the magnetic resonance appears at a shorter wavelength than the cutoff peak, the regions of $Re[\mu_{\text{eff}}]$ appear where $Re[\epsilon_{\text{eff}}]$ is positive, leading to positive values for $\text{Re}[n_{\text{eff}}]$.

The behavior of the effective optical parameters is very similar for a silver-DF structure. Panel (b) of Fig. 3.3 shows the same quantities as in panel (a) but now for a DF structure made of silver with the same geometrical parameters as in Fig. 3.3a, and $d = 600$ nm. As in the PEC case, the link between the condition $Re[\epsilon_{\text{eff}}] = 0$ and the cutoff wavelength is clearly observed. Moreover, a band of negative $Re[\mu_{\text{eff}}]$ presenting a smooth linear dependence with a_y also appears. The disappearance of this band for small a_y is due to absorption (calculations for a *lossless* silver show regions of negative $Re[\mu_{\text{eff}}]$ for small a_y). Absorption also explains why, as a difference with a PEC-DF, the region of negative $Re[n_{\text{eff}}]$ is larger than the one in which both $Re[\epsilon_{\text{eff}}]$ and $Re[\mu_{\text{eff}}]$ are negative. Note that the general condition for having $Re[n_{\text{eff}}] < 0$ is $Re[\epsilon_{\text{eff}}]|\mu_{\text{eff}}| + Re[\mu_{\text{eff}}]|\epsilon_{\text{eff}}| < 0$ (for more details see Ref. [23]). Notice however that the figure-of-merit (FOM), defined as the modulus of the ratio between the real and imaginary parts of n_{eff}, is maximum inside the region in which both $Re[\mu_{\text{eff}}]$ and $Re[\epsilon_{\text{eff}}]$ are negative.

The physical origin of the magnetic resonance described above clearly emerges when looking at the corresponding E-field patterns. In Fig. 3.4 we plot the E-field amplitudes for the DF structure evaluated at a xy plane between the two metal films [panels (a) and (b)] and at a xz plane that cuts the holes through their centers [panels (c) and (d)]. Panels (a) and (c) correspond to the PEC-DF case whereas (b) and (d) present the results for silver-DF. In these calculations, the long edge of the holes is fixed at $a_y = 0.5d$ and the E-fields are evaluated at the wavelength of the magnetic resonance. In the two frequency ranges analyzed (PEC and optical regimes), the E-field is mainly concentrated in the gap region between the metallic films and

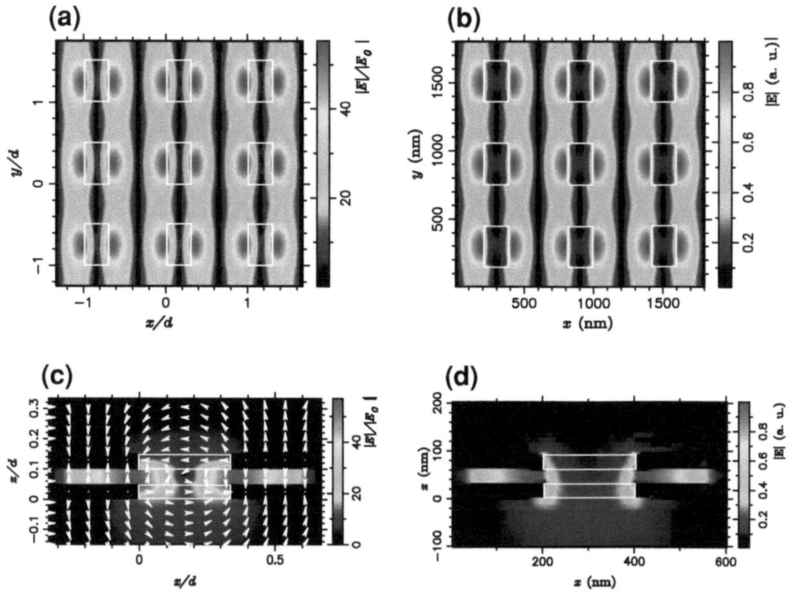

Fig. 3.4 **a** and **b** Render the electric field amplitudes at the magnetic resonance evaluated in the *xy* plane just in the middle of the dielectric slab ($z = 0.075\ d$). **c** and **d** Amplitudes in a *xz* plane ($y = 0.25\ d$). The geometrical parameters are as in Fig. 3.3 with $a_y = 0.5\ d$. **a** and **c** Correspond to the PEC case whereas **b** and **d** are for the silver-DF structure with $d = 600$ nm. *White lines* mark the positions of the holes

also has a strong standing wave character in the *x*-direction. The E-field intensity maxima are along the ridges of the holes, suggesting that in both cases two SPP-like modes that counter-propagate in the *x* direction are involved in the formation of the resonance.

A dielectric gap placed between two metallic films supports the propagation of SPP-modes that are usually called gap-SPPs [24]. For two non-perforated silver films of thickness $h_m = 30$ nm and separated by a 30 nm-vacuum gap (geometrical parameters of the silver-DF structure), the spectral location of the gap-SPP mode that could be excited by a normal incident plane wave can be calculated by evaluating the gap-SPP dispersion relation at $k_x = 2\pi/d$. This calculation leads to $\lambda_{\text{res}} = 1050\,\text{nm} = 1.75d$, that nicely coincides with the limit $a_y \rightarrow 0$ of the magnetic resonance leading to negative n_{eff} displayed in Fig. 3.3b. Folding of the gap-SPP dispersion relation at $k_x = 4\pi/d$ originates the small NRI-region located at $\lambda \approx 1.3d$ for small a_y observed in Fig. 3.3b. When the metal is a PEC, gap surface EM modes that are very similar to the gap-SPP modes in the optical regime are created due to the presence of a hole array drilled in the PEC films, even though non-perforated PEC films do not support the propagation of gap-SPP modes. The dispersion relation of these geometry-induced gap-SPP modes lies very close to the light line, explaining why the magnetic resonance band for a PEC-DF appears very close to $\lambda = d$ (see

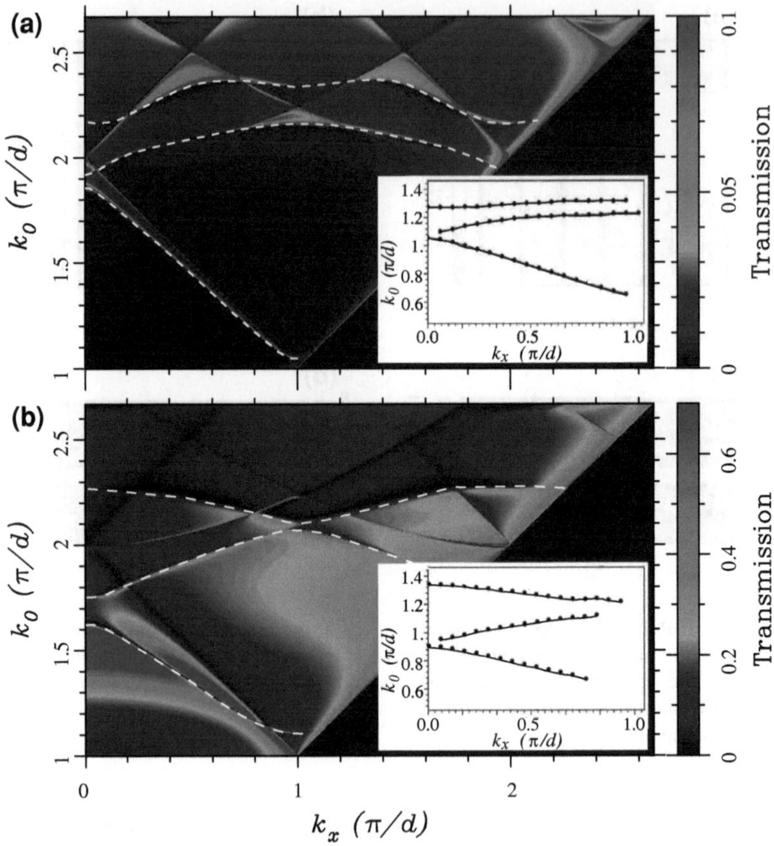

Fig. 3.5 Transmittance as a function of both incident wavenumber (k_0) and parallel momentum (k_x) for $a_y = a_x = 0.33\ d$ **a** and $a_y = 2a_x = 0.66\ d$ **b**. *White dashed* lines show the spectral locations of the gap-SPP modes in the holey structure.*Insets* corresponding dispersion relations for the gap-SPP modes in DF structures made of silver with $d = 600$ nm

Fig. 3.3a). The connection between the resonant magnetic response and the excitation of gap-SPP modes is highlighted for the PEC case in Fig. 3.4c. The electric field is pointing at opposite directions at the two sides of the dielectric slab, generating a displacement current that resembles that of the one created by a magnetic dipole parallel to the *y*-direction. This is the standard explanation of the magnetic behavior in DF structures, as described in Ref. [8]. But interestingly there is a link between the resonant magnetic response of a DF structure with the excitation of gap-SPP modes, both in the PEC and optical regimes.

If gap-SPP modes are involved in the resonant magnetic response, it is expected that this response will be very sensitive to the angle of incidence. This is demonstrated in the main panels of Fig. 3.5 that renders contour-plots of the transmittance versus wavenumber ($k_0 = 2\pi/\lambda$) and parallel momentum (k_x) for two different PEC-DF

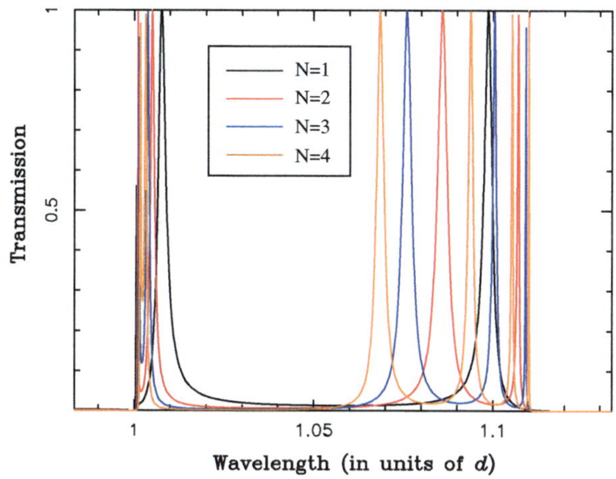

Fig. 3.6 Transmittance versus wavelength for DF-based structures in which the number of air gaps, *N*, is increased from 1 to 4. Here $a_x = a_y = 0.33d$ and $\epsilon_d = 1.0$. From Ref. [18]

structures. In panel (a), $a_y = 0.33d$ whereas panel (b) shows the case $a_y = 0.66d$. The spectral locations of the transmission resonances due to the excitation of gap-SPP like modes (leading to negative μ_{eff}) are underlined with white dashed lines. Due to the folding of the dispersion curves of the gap SPP-like modes inside the first Brillouin zone, there are three branches that present resonant magnetic response. For small holes, these curves highly disperse with k_x (angle of incidence), specially the lower branch. As a_y is increased (see panel (b)), the hybridization of the gap SPP-like modes with the cutoff resonance decreases the dispersion of the gap SPP-like modes with k_x. Insets in Fig. 3.5 show the corresponding results for the silver-DF structures. The locations of the optical gap-SPP modes (and their associated NRI) also disperse with parallel momentum, although this dispersion is less pronounced than the one found in the PEC case.

3.2.3 3D Metamaterials: Stacked DF Structures

The fact that DF structures display NRI behavior at near-infrared and optical frequencies is quite interesting but, however, they are not truly 3D metamaterials as their thicknesses are much smaller than the operating wavelength [25]. Therefore, the question on what happens to the NRI behavior of DF-based structures when many metal-dielectric-metal stacks are added is timely and important from both the fundamental and applied points of view. In what follows, we address this issue by making use of the CMM in which the inclusion of many multilayers is straightforward. In Fig. 3.6 we plot the evolution of the transmission spectra as the number of dielectric layers, *N*, is increased from 1 (the case we have analyzed up to now) to 4. The geometrical parameters of the 2D hole array we have used in these simulations are $a_x = a_y = 0.33d$ and $\epsilon_d = 1.0$. The metal is approximated as a PEC. As clearly

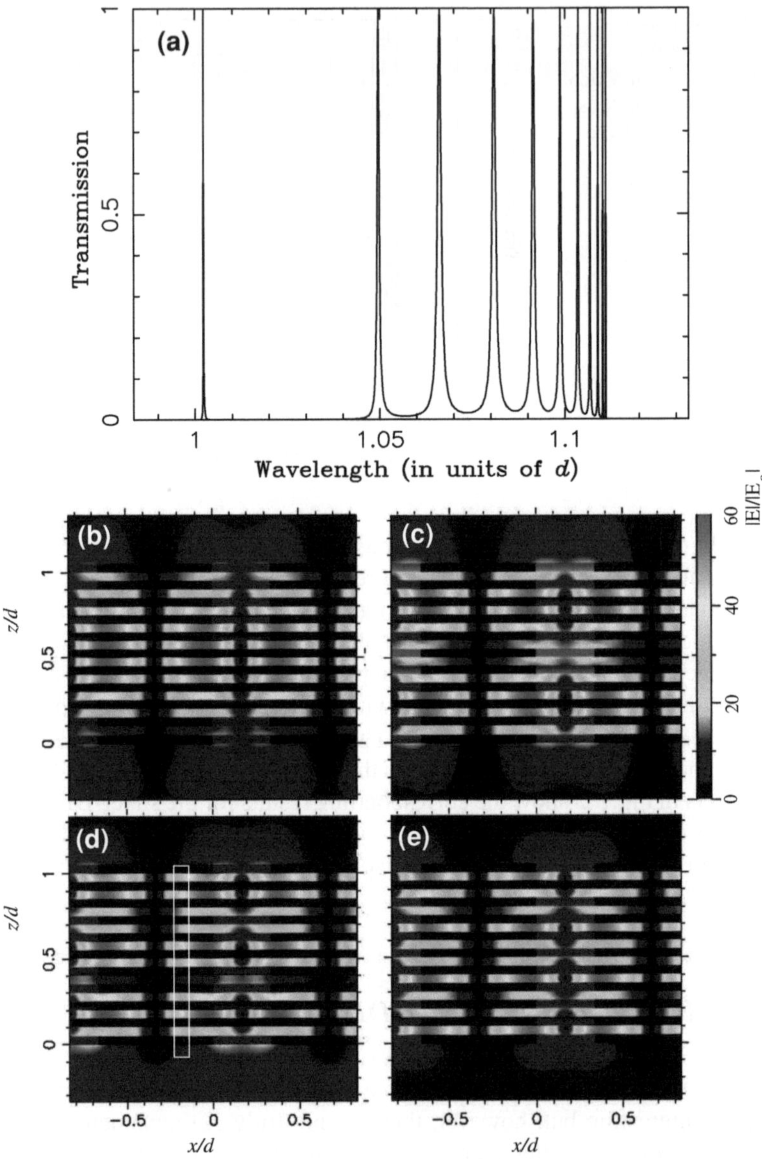

Fig. 3.7 **a** Transmittance versus wavelength for the case $N = 10$. The geometrical parameters are the same as the ones used in previous cases. **b–e** Display the amplitude of electric field patterns for **b** $\lambda = 1.049\,d$ (1st peak), **c** $\lambda = 1.066\,d$ (2nd peak), **d** $\lambda = 1.081\,d$ (3nd peak) and **e** $\lambda = 1.092\,d$ (4th peak) evaluated in a xz plane that cuts the holes passing through their centers. From Ref. [18]

seen in this figure, the inclusion of more and more layers results in the appearance of additional resonant features in the transmission spectrum. The number of these new

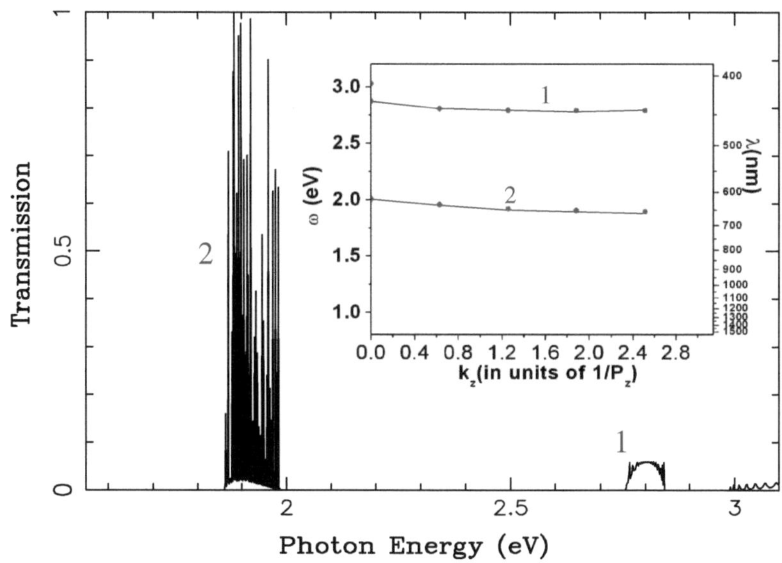

Fig. 3.8 Transmission versus photon energy (in eV) for a multilayered DF-based structure in which the number of air gaps is $N = 200$. The 200 peaks within the energy window $1.87 - 1.98$eV should reach 100 % transmission. The inset shows the dispersion relations (energy versus momentum in the z-direction) for the two NRI-bands as calculated with the FDTD method. From Ref. [18]

features exactly coincides with the number of dielectric layers, N. Associated with these transmission peaks, resonant behaviors of the magnetic response, μ_{eff}, leading to negative values are also observed. Note that the EOT peaks associated with the two surface EM modes of the entrance and exit surfaces of the structure are still present in the transmission spectrum but their linewidths are strongly reduced as N is increased. The fact that the number of transmission peaks coincides with N suggests that the origin of the multiple transmission peaks stems from the electromagnetic coupling between the spoof gap-SPP modes running at the air gap regions. These localized modes are electromagnetically connected via the 2D hole arrays of the PEC layers. This hypothesis is nicely corroborated in Fig. 3.7 which shows the E-field amplitude patterns (evaluated at a xz plane that passes through the center of the holes) for the first four resonances appearing for the case $N = 10$.

These patterns look like the different waveguide modes appearing in a Fabry-Perot cavity. They present standing wave character both in the x-direction (coming from the the interference between two counter-propagating gap SPP-like modes) and in the z direction, typical of Fabry-Perot like resonance. The first resonance (appearing for this set of parameters case at $\lambda = 1.049d$) is the fundamental mode in which no nodes are present in the z-direction. As the wavelength is increased (see panels (c)–(e)), more and more nodes emerge in the pattern. These four panels highlight the collective nature of the EM modes involved in the NRI response of multilayered

DF-structures. In the structure analyzed in Fig. 3.7 ($N = 10$), the thickness of the whole structure is still of the order of the wavelength.

The final system we want to present is a truly DF-based 3D metamaterial made of silver. In Fig. 3.8 we render the transmission spectrum (in this case transmission versus photon energy in eV) for a DF-based structure with $N = 200$ air gaps. The geometrical parameters are the same as in previous cases and we have used $d = 600$ nm as the period of the hole array. As expected, many transmission peaks located within a very narrow energy range emerge in the spectrum. The linewidth of this NRI band (marked with label 2 in Fig. 3.8) is of the order of 50 nm.

Interestingly, there is another NRI band (marked with label 1) appearing at higher frequencies. These results are corroborated by FDTD calculations on the infinite DF-structure. The inset of Fig. 3.8 displays the dispersion relation (frequency versus momentum in the z-direction) of these NRI bands as calculated with the FDTD method. There is an excellent agreement between the FDTD and modal expansion results that shows again the reliability of our theoretical framework. More importantly, these results demonstrate that the NRI behavior in DF structures is maintained as the number of DF-layers is increased. Note that in real metals in the optical or near-infrared regimes, the presence of absorption within the metal layer would limit strongly the NRI response in multilayered DF-based structures.

3.3 Conclusions

In conclusion, we have presented a complete theory of the NRI response of double-fishnet structures by analyzing two very distinct frequency regimes. Our results show that these structures can be interpreted as holey plasmonic metamaterials. Their electric permittivity is governed by the cutoff frequency of the hole waveguide. Negative values of the magnetic permeability are associated with the excitation of gap-SPP modes in the dielectric film. Finally, we have also analyzed how the negative refractive index response evolves when many double-fishnet units are stacked together. Multiple magnetic resonances emerge in these structures originated from the electromagnetic coupling between the different gap surface modes of the dielectric gaps. These results show the negative refractive index behavior is maintained in a truly 3D DF-based metamaterial.

References

1. V. Veselago, Sov. Phys. Usp. **10**, 509 (1968)
2. J.B. Pendry, Phys. Rev. Lett. **85**, 3966 (2000)
3. U. Leonhardt, Science **312**, 1777 (2006)
4. J.B. Pendry, D. Schurig, D.R. Smith, Science **312**, 1780 (2006)
5. D. Schurig, J.J. Mock, B.J. Justice, S.A. Cummer, J.B. Pendry, A.F. Starr, D.R. Smith, Science **314**, 977 (2006)

6. V. Veselago, L. Braginsky, V. Shklover, C. Hafner, J. Comput. Theor. Nanosci. **3**, 1 (2006)
7. J. Pendry, A. Holden, D. Robbins, W. Stewart, IEEE Trans. Microw. Theory Tech. **47**, 2075 (1999)
8. V.M. Shalaev, Nat. Photon. **1**, 41 (2007)
9. S. Zhang, W. Fan, N.C. Panoiu, R.M. Osgood, S.R.J. Brueck, Phys. Rev. Lett. **95**, 137404 (2005)
10. S. Zhang, W. Fan, K.J. Malloy, S.R.J. Brueck, N.C. Panoiu, R.M. Osgood, Opt. Express **13**, 4922 (2005)
11. C.M. Soukoulis, S. Linden, M. Wegener, Science **315**(5808), 47 (2007)
12. G. Dolling, M. Wegener, A. Schädle, S. Burger, S. Linden, Appl. Phys. Lett. **89**, 231118 (2006)
13. G. Dolling, C. Enkrich, M. Wegener, C.M. Soukoulis, S. Linden, Science **312**, 892 (2006)
14. M. Beruete, M. Sorolla, I. Campillo, Opt. Express **14**, 5445 (2006)
15. M. Kafesaki, I. Tsiapa, N. Katsarakis, T. Koschny, C.M. Soukoulis, E.N. Economou, Phys. Rev. B **75**, 235114 (2007)
16. J. Valentine, S. Zhang, T. Zentgraf, E. Ulin-Avila, D.A. Genov, G. Bartal, X. Zhang, Nature **455**, 376 (2008)
17. A. Mary, S.G. Rodrigo, F.J. Garcia-Vidal, L. Martin-Moreno, Phys. Rev. Lett. **101**, 103902 (2008)
18. A. Mary, S.G. Rodrigo, L. Martin-Moreno, F.J. Garcia-Vidal, J. Phys.: Condens. Matter **20**, 304215 (2008)
19. L. Martín-Moreno, F.J. García-Vidal, H.J. Lezec, K.M. Pellerin, T. Thio, J.B. Pendry, T.W. Ebbesen, Phys. Rev. Lett. **86**, 1114 (2001)
20. J.B. Pendry, L. Martín-Moreno, F.J. García-Vidal, Science **305**(5685), 847 (2004)
21. L. Martin-Moreno, F.J. García-Vídal, J.Phys.: Condens. Matter **20**, 304214 (2008)
22. D.R. Smith, S. Schultz, P. Markoŝ, C.M. Soukoulis, Phys. Rev. B **65**, 195104 (2002)
23. R.A. Depine, A. Lakhtakia, Microw. Opt. Technol. Lett. **41**, 315 (2004)
24. V. Economou, Phys. Rev. **182**, 539 (1969)
25. G. Dolling, M. Wegener, S. Linden, Opt. Lett. **32**, 551 (2007)

Chapter 4
Plasmonic Devices

4.1 Introduction

The emerging field of Plasmonics is based on exploiting the coupling between light and Surface Plasmon Polaritons (SPPs). Because of SPP modes are not constrained by the optical diffraction limit, it is hoped they could enable the construction of ultra-compact optical components [1–3]. Among these components would have SPP sources and waveguides for sending information through, from one place to another on a hypothetical Plasmonic-chip.

During this thesis, we have been working on these very interesting topics in the potential use of SPPs as we will see below:

a. The first topic is addressed in Sect. 4.2, which is related with the inherent difficulty of exciting SPPs in an efficient manner [4, 5]. Given that SPPs cannot be directly excited due to the mismatch momentum with freely propagating light, the light-plasmon coupling efficiency becomes of crucial importance for the success of any plasmonic device. In these works, we investigated a nano-structured metallic system that allows SPPs being unidirectionally excited from the source, in this case, a slit drilled through a metal film. It is vital for the realization of SPPs nano-optical components that the relatively poor light-SPP coupling is improved. Another difficulty in plasmonic circuits is that the incident light, which is conventionally used to launch SPPs in a metal film [6–9], is a significant source of noise, unless directed away from a region of interest which then decreases the signal and increases the system's size. Back-side illumination of subwavelength apertures in optically thick metal films [10–16] eliminates this problem but does not ensure a unique propagation direction for the SPP.
   In Sect. 4.2 we present a novel back-side slit-illumination method which incorporates a periodic array of grooves carved into the front side of a thick metal film. Bragg reflection enhances the propagation of SPPs away from the array, enabling them to be unidirectionally launched from, and even focused to, a localized point. We also show a comprehensive study on the modulation (enhancement or

Springer Theses, DOI: 10.1007/978-3-642-23085-1_4,

suppression) of such coupling efficiency. Our approach is based on simple wave interference and enables us to make quantitative predictions which have been both numerically and experimentally confirmed at both the near infrared and telecom ranges. From the theoretical standpoint we use the finite-difference time-domain (FDTD) method (Sect. 1.2) and a Coupled Mode Method (CMM)(Sect. 1.3) version adapted for investigating the optical response of finite set of 1D-indentations. The CMM method has been conducted by Dr. F. López-Tejeira at the Departamento de Física de la Materia Condensada (Universidad de Zaragoza).

b. In Sect 4.3 the second topic is investigated, which relates with our ability to control light on a surface once it has been "launched". In references [17–19] we presented a theoretical study on two different proposals for guiding EM fields, namely the modes supported by a carved triangular groove in metal and, in a way, the "complementary" structure, a triangular metal wedge. This challenging matter, the guiding of light within a subwavelength cross section, is especially compelling due to the ever increasing demands for miniaturization of photonic circuits. Therefore, the realization of subwavelength guiding structures is a key factor for miniaturization, because these components would permit denser waveguide packaging without crosstalk, and lower waveguide bending loss.

   Light may be confined in the direction perpendicular to a flat metallic surface at energies below the metal plasma frequency if it couples to SPPs. Various geometries have been proposed to achieve confinement of the plasmon-polariton in the plane transverse to the propagation direction [20–25]. Among these proposals, the plasmon-polariton guided by a V-shaped groove carved in a metal (channel plasmon-polaritons, CPPs) and the modes supported by a metallic wedge (wedge plasmon-polaritons, WPPs) are particularly interesting. CPPs were theoretically suggested by Maradudin et al. [26] and subsequently studied in the visible regime [22, 27]. CPPs have been experimentally investigated at telecom wavelengths [28], displaying strong confinement, low damping, and robustness against channel bending. Thanks to these properties, prototypes of basic devices could be demonstrated [29]. On the other hand, WPPs were also shown to support strongly localized plasmons, which has been demonstrated both theoretically [18, 23] and experimentally [23, 30].The first devices were developed with the help of effective index approximations. An effective index approximation can deliver information about the dispersion relation, but it is expected to be inaccurate for frequencies close to the mode cutoff and is unable to determine modal shape and polarization. The functionality of many devices relies on the overlapping of electromagnetic fields at various sites inside the device. For this reason the knowledge of the modal shape is essential to provide a solid foundation for the design of CPP-based devices.

   In Sect. 4.3, we present rigorous simulations of guided CPPs and WPPs aimed at elucidating their characteristics at telecom wavelengths, including full vectorial modes, dispersion, and losses. The simulations were performed with two rigorous electrodynamic techniques: the FDTD method (Sect. 1.2) and the multiple multipole method (MMP) [31], this latter carried out by Dr. E. Moreno at the Departamento de Física Teórica de la Materia Condensada, Universidad Autónoma de

Madrid. Our goal thus is to understand the fundamental CPP and WPP modes guided by realistic structures at telecom wavelengths [28]. Nevertheless, in order to comprehend the behavior in this regime, which is close to cutoff, we will consider a broader spectrum, higher order modes, and a number of different geometries. Finally, we design optical devices which rest on the results previously found. In this way, we propose a WPP to SPP geometry-driven conversion device. Besides, it is demonstrated both theoretically and experimentally that "tapered" CPP supporting structures can enhance EM fields near the surface.

4.2 An Efficient Source for Surface Plasmons

Because of the so-called "excess of momentum" with respect to light of the same frequency, SPPs cannot propagate away from a planar surface and are thus bound to and guided by it. As a consequence of such binding, SPP modes can be laterally confined below the diffraction limit, which has raised the prospect of SPP-based photonic circuits [1, 2, 32]. To build up this kind of circuits one would require a variety of components in which incident light would be first converted in SPPs, propagating and interacting with different devices before being recovered as freely propagating light. Hence, a great deal of attention has been devoted to the creation of optical elements for SPPs [29, 33–37], as well as to the efficient coupling of freely-propagating light into and out of them. This latter issue constitutes the fundamental bottleneck that must be overcome in order to fully exploit the potential of SPPs, given that established techniques for SPP generation (which make use of prism [6, 7], grating [8] or nanodefect [9] coupling) require that the system's size be well out of the subwavelength scale in order to obtain a neat SPP signal. On the other hand, *p*-polarized back-side illumination of subwavelength apertures in optically thick metal films [10–16, 38] prevents both damping and signal blinding but it does not ensure only a propagation direction for the generated SPPs.

In the present section we present a back-side slit-illumination method based on drilling a periodic array of indentations at one side of the slit. It will be demonstrated that the SPP beam emerging from the slit to its corrugated side can be back-scattered in such a way that it interferes constructively with the one propagating in the opposite direction, thus obtaining a localized unidirectional SPP source. We provide a comprehensive version of such proposal and discuss in some extent its range of validity.

4.2.1 Description of the Proposal

A picture of the proposed structure is shown in Fig. 4.1. The starting point for such a design can be found in a previous work on 1D SPP scattering by means of the CMM [39, 40]. In order to cope with SPP launching, it was considered a single slit flanked by

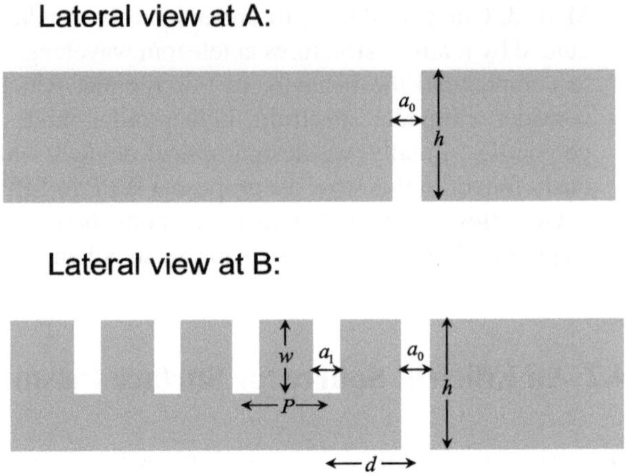

Fig. 4.1 Parameters $\{a_i, h, w, d, P\}$ defining the geometry of the system are also shown

an array of indentations (rectangular grooves) placed in the output surface of a thick metallic film. Eventually, the distance between the slit and indentations was taken to be infinity. In this way, the slit merely played the role of a theorist's SPP-launcher, as far as it can be shown that the field created by the slit corresponds to SPP illumination into the grooves. Besides, it was also found a simple geometrical condition for the groove array to behave as a perfect Bragg mirror, associated to the low-λ edge of the plasmonic band gap for the periodic system. For narrow subwavelength indentations, the spectral locations of these edges can be approximated by folding the dispersion relation of SPPs for a flat metal surface into the first Brillouin zone [41]. Such folding results in

$$k_p P = k_0 Re[q_p] P = m\pi, \quad m = 1, 2 \ldots \tag{4.1}$$

where P is the period of array and k_p holds for in-plane plasmon wave-vector.

Combining these two elements, one can obtain a remarkably simple scheme for the modulation of SPP coupling-in: given an incident wavelength, let us place at a distance d from one of its sides a groove array for which reflectance rises to a maximum. Hence, any SPP emerging from the slit will be mainly back-scattered and interfere either constructively or destructively with the one leaving the slit by its opposite side. This interference can be tuned by adjusting the separation d between the slit and the first groove of the array, defined centre to centre. The total phase difference, ϕ, between the interfering SPPs will then consist of the phase change upon reflection plus the additional shift resulting from the two different path lengths along the metal:

$$\phi = \phi_R + 2Re[k_p]d, \tag{4.2}$$

According to Eq. 4.2, constructive or destructive interference should occur for those phase values equal to, respectively, even or odd multiples of π.

We have found that ϕ_R is close to π over a wide range of groove depths for $a/\lambda \leq 0.2$ at both NIR and telecom ranges, as stated in Ref. [5]. Taking this result into account and substituting for k_p from Eqs. 4.1 into 4.2 yields

$$\phi(\lambda_R) = \left(\frac{2md}{P+1}\right)\pi, \tag{4.3}$$

which reduces the design of our proposed scheme to a suitable choice of the d/P ratio.

However, the key point of the proposal still relies on the properties of SPPs. The EM fields radiated by the slit cannot be considered "purely plasmonic" at distances smaller than several wavelengths, where the field consist partly of a different wave that decays as $x^{-1/2}$, the so-called "creeping wave" [39, 42]. This makes necessary detailed computations that go beyond the simple model described before. On the other hand, SPPs have been thought to dominate the EM field beyond this region. However, irrespectively of the metal considered, the long-distance asymptotic limit of the EM field at metal surface is not the SPP but a different type of wave known as Norton wave (NW) [43]. This wave decays as $x^{-3/2}$ for 2D-dipole sources and as x^{-2} for 3D-dipoles, a behavior substantially different from the exponential decay of SPPs. Note nevertheless that contribution to the surface field of NWs is negligible at distances smaller than ~5 times the corresponding SPP propagation length [43]. At the near infrared and telecom the propagation length of a SPP is greater than 100 microns by far, so NWs do not play a significant role.

In order to characterize the efficiency of the slit + array system as a SPP-launcher for any slit-to-array separation, we introduce its "efficiency ratio", E_R: given that the array be located at the left side of the slit (see Fig. 4.1), E_R is defined as the quotient between the current intensity of right-propagating SPP with and without the grooves. Strictly speaking, E_R provides the efficiency of the output side of the device. The total efficiency, defined as the percentage of incident energy transferred onto the plasmon channel, strongly depends on the illuminating setup. E_R should vary within the interval $[0, 4]$ showing a dependence on the distance between the illuminating slit and the groove array. More importantly, $E_R > 2$ implies that the right-propagating SPP current in the presence of grooves is larger than the total SPP current (left- plus right-moving) in the single slit case, so some of the power radiated out of plane is redirected onto the SPP channel. According to this simple wave interference model,

$$E_R \approx |1 + re^{2ik_p d}|^2, \tag{4.4}$$

where r is the complex reflection coefficient of the groove array for SPPs. To check the validity of Eqs. 4.3 and 4.4 for slit-to-array separations outside the asymptotic regime, we have carried out numerical calculations of EM fields by means of both CMM (Sect. 1.3) and FDTD (Sect. 1.2). The system under consideration is intended to operate at a wavelength of 800 nm on a gold film (Table 1.1). We consider an array of ten grooves with a period $P = 390$ nm. The depth of the grooves is chosen to be $w = 100$ nm, while the width of both grooves and slit is $a = 160$ nm, which are typical experimental parameters. Figure 4.2a shows the comparison between Eq. 4.4 and numerical evaluations of E_R, as well as the location of interference maxima (vertical lines) predicted by Eq. 4.3 for $m = 1$. The agreement between CMM and FDTD results is excellent, except for very small distances ($d \approx 2a$) between the slit and the array, when intra-wall coupling between the slit and the first groove has to be taken into account. As can be seen, the locations of maximum E_R are accurately predicted by Eq. 4.3, which allows us to design SPP-launchers without elaborate numerical calculations. Moreover, the simplified model of Eq. 4.4 provides a good approximation to E_R with the sole input of r. This also implies that non-plasmonic contributions to groove illumination play a minor role in the occurrence of either constructive or destructive interference, for this particular structure choice, which is clearly described by Eq. 4.4 with the except of minor shifts.

In addition to the efficiency ratio, field patterns in both minimum and maximum condition were also calculated using the FDTD method. As shown on Fig. 4.2b, SPPs are completely absent from the left side of the slit whereas field intensity at its right side is clearly modulated by the slit-to-array separation, which also governs the spatial distribution of the field that is radiated into the vacuum.

4.2.2 *Results*

Near-Infrared

To test experimentally the proposal, several samples were prepared at Laboratoire de Nanostructures, ISIS, Universitè Louis Pasteur, in the group of Prof. T.W. Ebbesen. Using a focused ion beam in 300-nm-thick gold films they made samples for different values of *d*, with all other geometrical parameters being the same as in the previous calculations. Each sample consists of a single long slit flanked by a finite periodic groove array that extends over only half of the slit length (see Fig. 4.3).

This sample design allows the quantitative experimental study of the SPP launching efficiency, as the "isolated" slit (upper part) can be used as an in-chip reference. The set of samples was imaged by a photon scanning tunnelling microscope at Laboratoire de Physique de l'Universitè de Bourgogne, (Dijon, France) in the group of Prof. A. Dereux, making use of an incident focused beam illumination for frequencies in the [765,800] nm interval. Owing to specific features of the experimental set-up used for measurements in the optical regime, the incident laser beam was directed on the sample (attached to a prism) under an angle of 43° with respect to the normal. However, it should be noted that the choice of angle of incidence is not critical for the spatial distribution of transmitted energy, as a subwavelength slit in an optically thick metal film transmits only in the fundamental mode. For each distance, *d*, a pair of images was recorded by scanning at a constant distance of about 60–80 nm from the sample surface. The first image of the pair, corresponding to the SPP launching by a single slit, was obtained by focusing the laser beam on the upper part of the slit. For the second image, the laser beam was moved to the

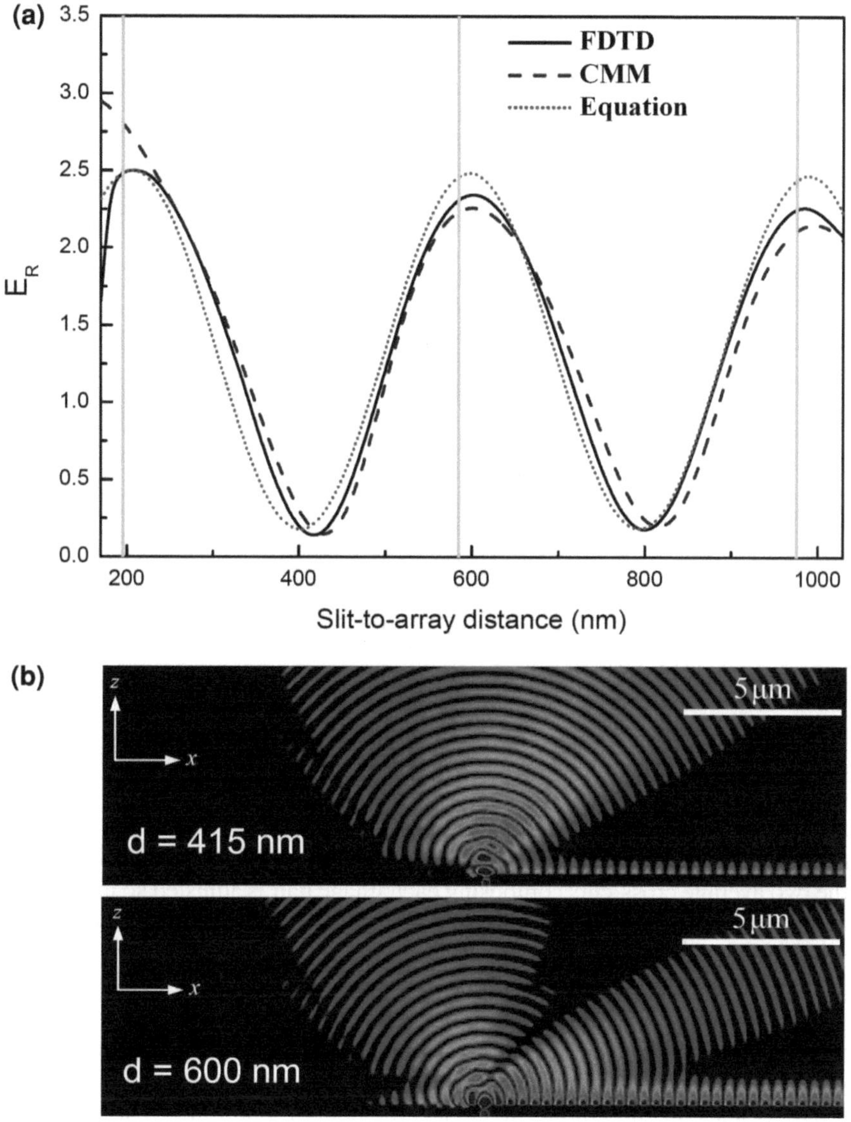

Fig. 4.2 Numerical results for the SPP launcher at wavelength $\lambda = 800$ nm. **a** Dependence of the efficiency ratio E_R on the slit-to-array distance. The geometrical parameters defining the system are: slit and groove widths $a = 160$ nm, groove depth $w = 100$ nm and array period $P = 390$ nm. The figure renders the curves obtained by means of FDTD (*solid*), CMM (*dashed*) and Eq. 4.4 (*short-dotted*). Vertical lines mark the positions of E_R maxima according to (4.3). **b** Calculated $|Re[H_y]|$ distributions over xz plane for two different distances corresponding to minimum and maximum values of E_R at $\lambda = 800$ nm

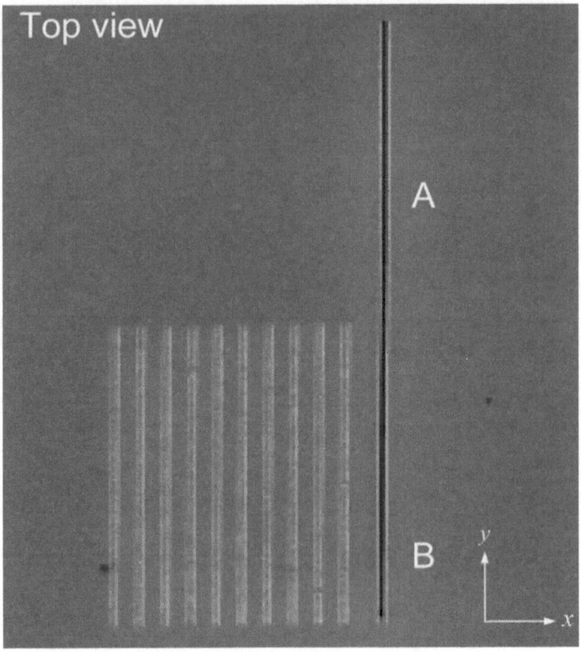

Fig. 4.3 Scanning electron micrograph and schematic diagrams of the proposed structure

lower part to collect the data for the slit+grating case. Image for $d = 585$ nm is shown in Fig. 4.4a. This figure clearly shows that the grating increases the intensity of the right-propagating SPP for $d = 585$ nm. To quantify this effect, an average longitudinal crosscut of each image is obtained by using 20 longitudinal cross-cuts, corresponding to different coordinates along the slit axis. Then, the relative position of the two average cross-cuts composing each image pair is adjusted so that the saturated areas (that is, the signal taken right on top of the slit) are superimposed. Finally, the experimental efficiency ratio, E_R, is extracted by averaging the ratio between the two curves along the longitudinal cross-cut. Figure 4.4b shows experimental results for E_R for the five different samples fabricated. The agreement between the experimental data and the FDTD predictions is quite remarkable (especially when taking into account that each experimental point corresponds to a different sample), showing that the presence of the grating modulates the coupling into the right-propagating SPP.

Telecom

Similar samples to those used in the NIR measurements were designed to operate at the telecom range by up-scaling the period of the array and its separation from the slit (see Fig. 4.5a). Telecom measurements were conduced by I.P. Radko and Prof. S.I. Bozhevolnyi in Aalborg (Denmark) at the Department of Physics and

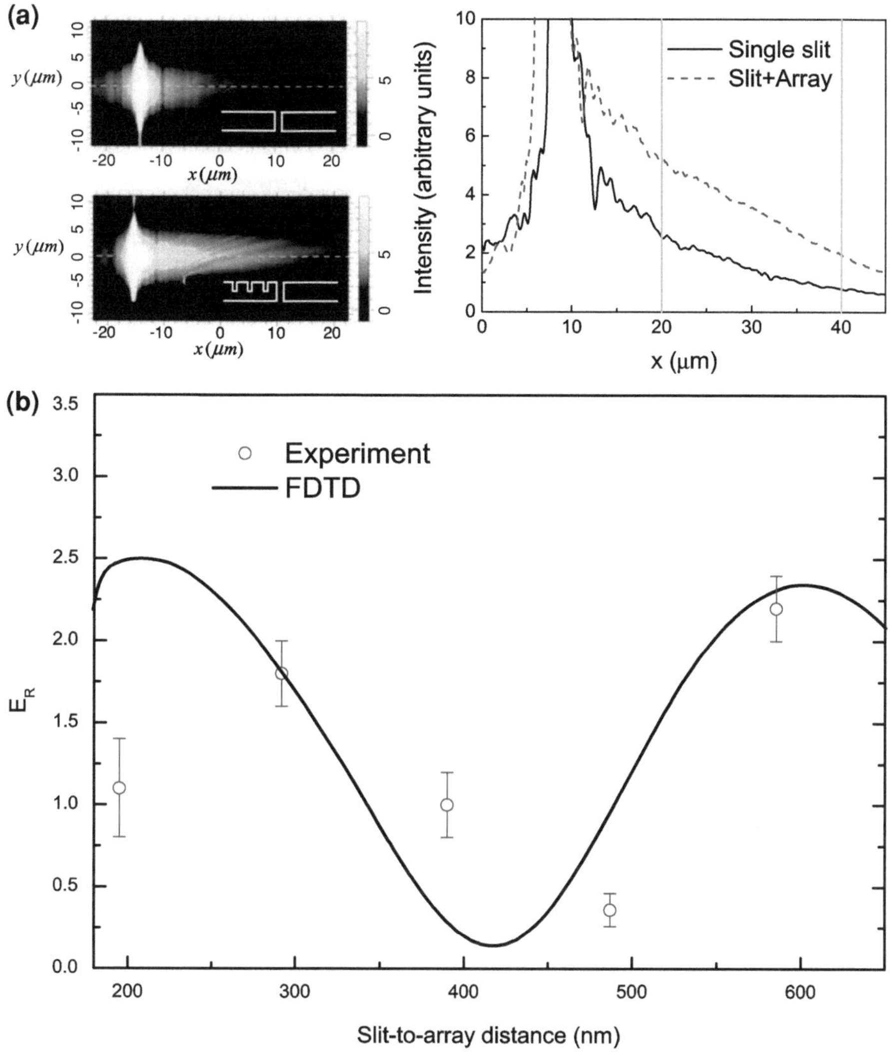

Fig. 4.4 Experimental measurement of E_R at $\lambda = 800$ nm for the same geometrical parameters as in Fig. 4.2. **a** Photon scanning tunneling microscope (PSTM) micrographs recorded for a sample with $d = 585$ nm at both "single slit" (*top*) and slit+array configurations (*bottom*). The *right* panel shows the two cross-cuts from which E_R is obtained. *Vertical lines* define the interval along the ratio is averaged. **b** Experimental (*circles*) and numerical (*solid line*) values of E_R as a function of slit-to-array distance. The error bars represent the standard deviation over a set of different structures with the same nominal parameters

Nanotechnology. During the experiments, they found an instability in the illumination setup that resulted in a noticeable variation of SPP intensity while taking place the near-field scan process, which took about 45 min per image. As a consequence of

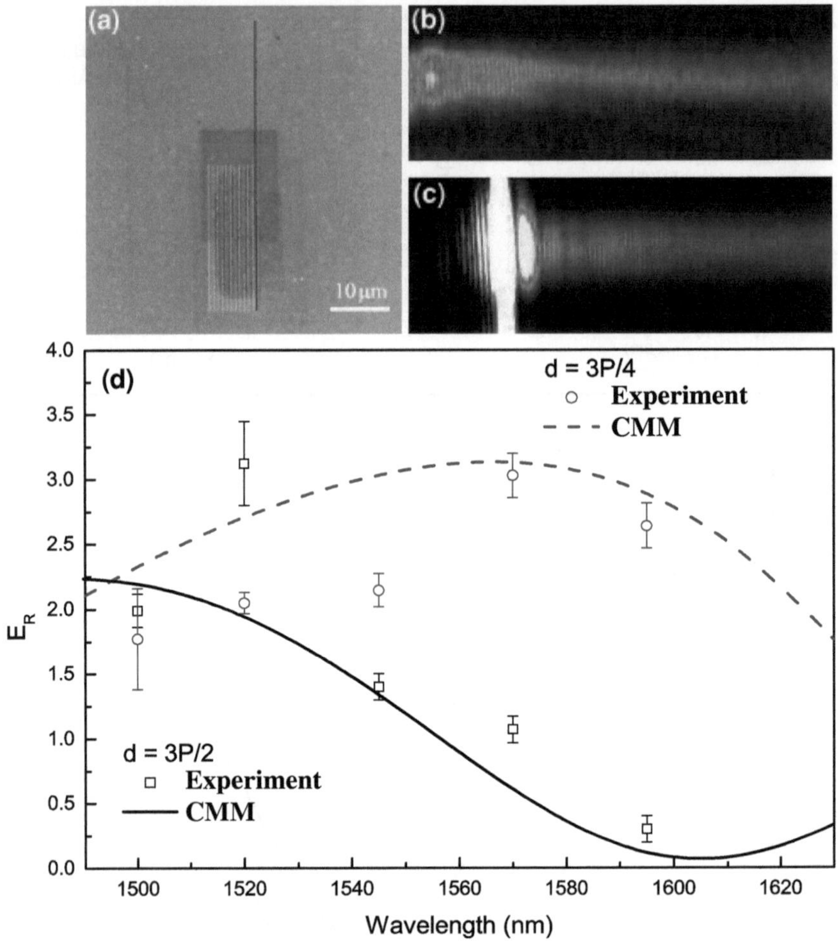

Fig. 4.5 Spectral dependence of E_R at the telecom range. **a** Scanning electron micrograph of the sample. The geometrical parameters are: slit length $L = 50\,\mu$ m, slit width $a_0 = 400$ nm, groove width $a = 200$ nm, groove depth $w = 100$ nm and array period $P = 750$ nm. **b** Near field image recorded with the laser beam focused at the "isolated slit" position of a sample with $d = 3P/2 = 562$ nm. (Size $= 70 \times 26\,\mu\text{m}^2$, $\lambda = 1{,}520$ nm). **c** Same for slit+array focusing. **d** Spectral dependence of E_R for slit-to-array distances of $d = 3P/2 = 1{,}125$ nm (experiment: *squares*; theory: *solid line*) and $d = 3P/4 = 562$ nm (experiment: *circles*; theory: *dashed line*)

those intensity jumps, the technique used to evaluate the "efficiency ratio" in the NIR became unsuitable. Instead, E_R was found as the SPP signal ratio taken from each pair of near-field images (with and without side grooves) at the same distance from the slit. This reference distance ($\approx$50 μm) is chosen so that the non-plasmonic field contribution can be disregarded, whereas the SPP signal is still substantial for the quantification. To decrease the uncertainty of the measured efficiency, a series of scans were performed for every structure and wavelength measurements, conducting independent adjustments, and the subsequent averaging of the E_R values obtained. Hence, the error E_R represents a statistically estimated deviation.

A typical pair of near-field optical images is presented in Fig. 4.5b and c. For telecom wavelengths, the SPP propagation length is increased up to $\approx$200 μm. Panel (c) features a strong SPP beam propagating away from the slit in the direction opposite to the array and thereby demonstrating unidirectional SPP excitation. Averaged results and estimated errors for E_R are rendered in Fig. 4.5d. Notice that the validity of our proposal is now tested in a different way: for a given slit-to-array separation, E_R is measured within the wavelength range 1,500–1,620 nm, so that the phase difference described by Eq. 4.2 changes as the wavelength increases, providing the conditions for constructive or destructive interference. This spectral dependence of the efficiency is different for different slit-to-array separations. For the case of the sample with $d = P + P/2 = 1{,}125$ nm, E_R decreases as the wavelength increases (with the only exception of the data point at 1,520 nm), evolving from a favorable regime ($E_R \approx 2$) to one in which coupling into SPPs is clearly diminished by the array ($E_R < 1$). Conversely, $E_R \approx 2$ for the sample with $d = 3P/4 = 562$ nm all over the range. As can be seen, the comparison between experiments and CMM is rather satisfactory.

Finally, we have to mention that the proposed approach for the excitation of localized unidirectional SPP beams can also be combined with the appropriate design modifications to create functional components for SPP focusing to a spot or tuning the SPP beam divergence. If $E_R \geq 2$ is expected for a given slit+array set, its circular bending may produce a converging gaussian beam whose waist length and radius can be adjusted changing the curvature. Several curved SPP focusers have been studied previously [44–48], but our scheme presents the advantage of preventing SPPs to escape in the opposite direction to the focus. Although the rigorous modeling of SPP coupling at curved structures is rather complicated, we expect Eq. 4.3 to still provide a good estimation for the proper design of the structure, for moderate curvatures. On that assumption, several samples were fabricated consisting of an arc-of-a-circle slit flanked by the corresponding array of parallel bent grooves (see Fig. 4.6a–c). Geometrical parameters a_0, a, w, P are the same as in Fig. 4.5, whereas slit-to-array distance is set to $d = 3P/2 = 1{,}125$ nm.

As shown in Fig. 4.6d–f, the effect of SPP launching and focusing can be appreciated already at the stage of far-field adjustment due to weak out-of-plane SPP scattering by surface roughness. Near-field images of SPP excitation on those structures recorded at free-space wavelength of 1,520 nm are presented in Fig. 4.7. These images clearly demonstrate the ability of a curved slit to excite a convergent SPP beam, this effect being sufficiently enhanced due to the side grooves (cf. [45, 46]). With the smallest radius of curvature (30 μm), focusing to a confined spot having size $3 \times 3\ \mu\text{m}^2$ is observed (see the cross cuts in the lower left panel of Fig. 4.7). The SPP beams excited on the less curved structures present an extended waist (Fig. 4.7b and c), which scales (at least visually) according to expectations, providing a wider, and hence less divergent, SPP beam. This might be useful for particular applications, e.g., in sensing of elongated biological samples or in coupling to low-numerical-aperture waveguides.

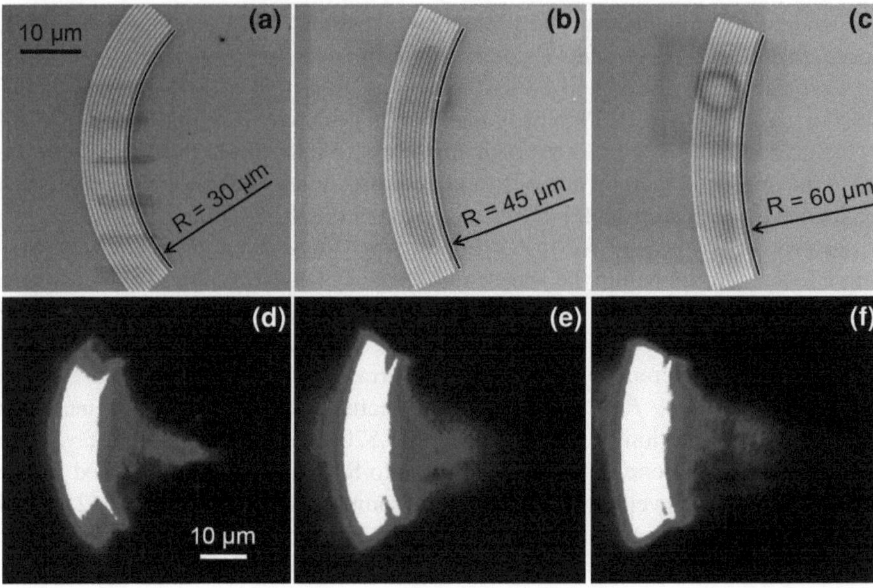

Fig. 4.6 **a** Scanning electron micrograph of the curved structure, characterized by slit and groove widths of 400–200 nm, respectively, groove periodicity $P = 750$ nm, groove depth $w = 100$ nm and slit-groove distance $d = 1{,}125$ nm. Film thickness $h = 280$ nm, curvature radius $R = 30\,\mu$m and slit chord length $L = 40\,\mu$m. **b, c** Same for $R = 45\,\mu$m and $R = 60\,\mu$m. **d–f** Far field images of SPPs excited on the structures (**a–c**), respectively, recorded with a charge-coupled device camera

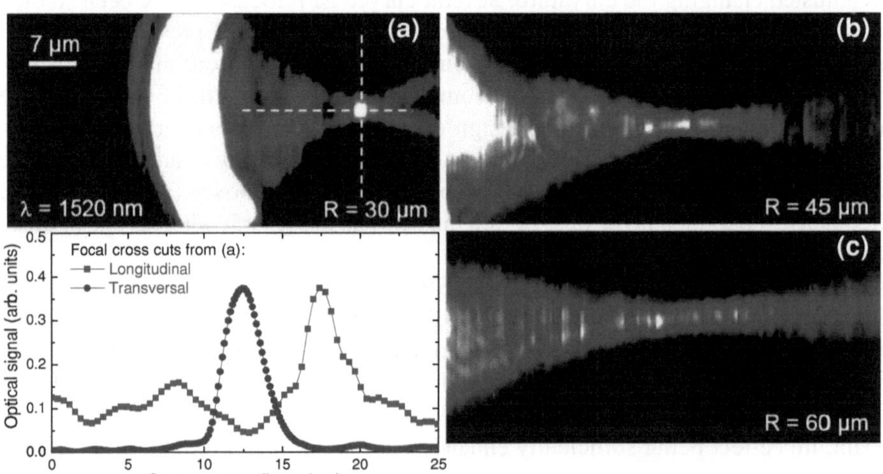

Fig. 4.7 **a–c** Near-field images (size $64 \times 32\,\mu m^2$) of SPPs excited on the structures in Fig. 4.6 at $\lambda = 1{,}520$ nm. Lower *left* panel depicts cross cuts obtained from (**a**) by dissecting the SPP focal spot along longitudinal and transversal directions

4.3 Guiding and Focusing EM Fields with CPPs and WPPs

As we said, CPPs are electromagnetic modes supported by grooves carved in metallic surfaces whereas WPPs are the corresponding modes sustained by metallic wedges. Roughly speaking, one could say that the electromagnetic field of CPP and WPP modes is guided along the bottom of the groove or the edge of the wedge, respectively.

Behavior of electromagnetic fields next to corners and edges has arisen the interest of scientific community since a long time. The first report dealing with these geometries within the context of surface plasmons was published by Maradudin et al. [49]. The mentioned paper considers an idealized geometry in the electrostatic approximation. The next landmark, in 2002, was a complete treatment including retardation for realistic geometries [26]. After this year the number of works, both theoretical and experimental, reporting on CPPs and WPPs has rapidly increased. We can mention the following reasons for this interest. First, the achievement of tightly confined modal fields and long propagation lengths count among the main design goals. We will see that CPPs and WPPs feature good confinement and a reasonable propagation length, and are therefore promising candidates. Second, the planar paradigm is preferred from a technological perspective, and the modes studied in this chapter fit well with planar metallic structures. From a more fundamental point of view, let us remark that edges and corners appear in other structures. In this sense CPPs and WPPs constitute building blocks that show up in other kinds of plasmonic guides such as stripes, trenches, gaps, and so on. Thus, understanding the properties of CPPs and WPPs is very useful for the design of another kind of plasmonic waveguides. Let us mention that, in spite of the difficulties to fabricate narrow angle CPPs and WPPs, these have been already demonstrated and CPP-based functional devices have been reported [50]. Recently, CPPs have been experimentally investigated at telecom wavelengths [28], displaying strong confinement, low damping, and robustness against channel bending. Thank to these properties, prototypes of basic devices have been demonstrated [29].

The mentioned devices have been developed and analyzed with the help of the effective index approximation. The effective index approximation can deliver information about the dispersion relation, but it is expected to be inaccurate for frequencies close to the mode cutoff and is unable to determine either the modal shape, the polarization and the propagation length, as we have said in the introduction.

As mentioned above, the goal of this section is to provide an understanding of the properties of CPP and WPP modes beyond the effective index approximation. We will describe the behavior of modal field, dispersion, modal size, propagation length, and the dependence of these magnitudes on various parameters by means of rigorous computational electrodynamics techniques (Fig. 4.8).

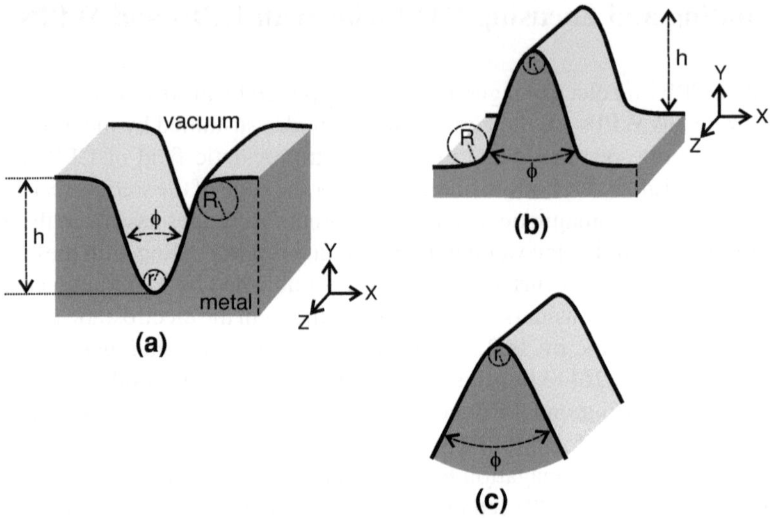

Fig. 4.8 Schematics of structures supporting channel and wedge plasmon polaritons. **a** Groove carved in surface. **b** Wedge grown on surface. **c** Infinite wedge

4.3.1 Channel Plasmon Polaritons

The functionality of many devices relies on the overlapping of electromagnetic fields at various sites inside the device. For this reason the knowledge of the modal shape is essential to provide a solid foundation for the design of CPP-based devices. Here we present rigorous simulations of guided CPPs aimed to elucidate their characteristics at telecom wavelengths, including full vectorial modes, dispersion, and losses.

Nevertheless, in order to comprehend the behavior in this regime, which is close to the waveguide cutoff, we will consider a broader spectrum, higher order modes, and a number of different geometries. The simulations have been performed with two rigorous electrodynamic techniques: MMP [31] and FDTD (Sect. 1.2). Within the MMP method the corners are rounded (10 nm radius of curvature). FDTD results were converged for a mesh of about 5 nm. Such fine meshes are essential, specially for wavelengths shorter than $\approx 0.8\,\mu$m. The grooves are carved in gold and we employ experimentally measured values (Table 1.1) of the dielectric permittivity ε.

Figure 4.9 shows the dispersion relation for a non-truncated groove with an angle of $\phi = 25°$ and infinitely long sides. This structure sustains two modes, being termed CPP(∞) (see right insets), which are outside the dispersion line of the SPP at a flat surface. The modal shape (time averaged electric field amplitude) is shown in the right insets for a wavelength of $\lambda = 0.6\,\mu$m. In the figure it is also plotted the dispersion relation for a non-truncated metallic wedge of angle $\phi = 102.5°$ and infinitely long sides. The corresponding wedge mode running along the edge is termed WPP(∞) (see left inset). WPP(∞) for this ϕ will be relevant when we later truncate the above groove at a finite height: it corresponds to the edges at both sides of the finite-height

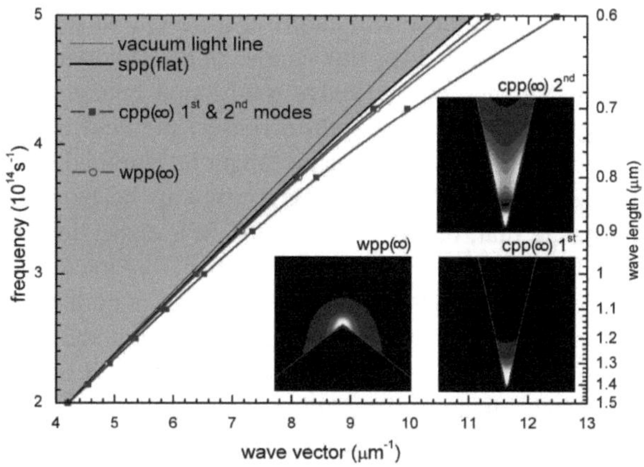

Fig. 4.9 Dispersion relation for various modes. *Black thick line* SPP mode on a flat surface. *Green lines* (*squares*) CPP(∞) modes for an infinitely deep groove. *Red line* (*open circles*) WPP(∞) mode for an infinitely deep wedge. *Right insets* time averaged electric field amplitude of the two CPP(∞) modes at 0.6 μm. *Left inset* same for the WPP(∞) mode. The lateral size of the *insets* is 2 μm

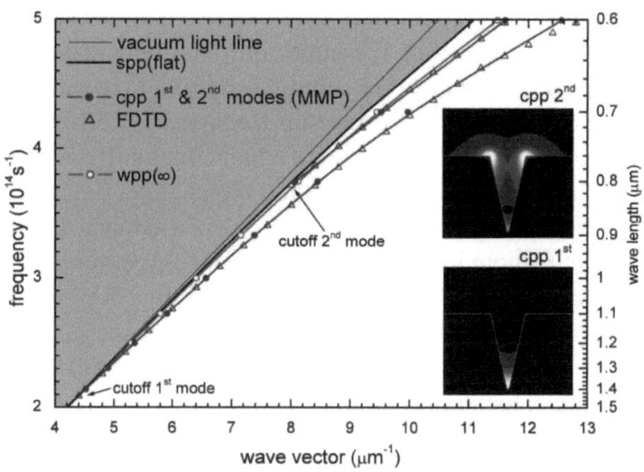

Fig. 4.10 Dispersion relation for various modes. *Black thick line* SPP mode on a flat surface. *Blue lines* (*full circles*) CPP modes for a groove of height 1.17 μm (computed with MMP method). *Triangles* same as before computed with FDTD method. *Red line* (*open circles*) WPP(∞) mode for an infinitely deep wedge. *Insets* time averaged electric field amplitude of the two CPP modes at 0.6 μm. The lateral size of the *insets* is 2 μm. From Ref. [17]

groove. The WPP(∞) modal field at 0.6 μm is shown in the left inset. For increasing wavelength all three modes approach the SPP line (none of them has a cutoff). In this process modal shapes remain qualitatively the same, the only difference being that the fields are expelled away from the groove or wedge corners.

Figure 4.10 represents a similar plot but now a groove of finite height is considered, being arbitrarily chosen 1.17 μm (however it is of the order of typical experimental values). The CPP modes exhibit now a cutoff at different wavelengths (~1.44 μm for the first mode and ~0.82 μm for the second one). This was advanced in Ref. [51], and it is a consequence of the above mentioned behavior of the fields for increasing wavelength. As the wavelength increases, the field is pushed out of the groove and, after a certain threshold, it can no longer be confined by the groove sides and is radiated in the form of SPPs along the contiguous horizontal metal surfaces. It is important to realize that, before reaching the SPP dispersion line, both modes approach and cross the WPP(∞) line. This means that close to cutoff the CPP modes must be hybridized with the modes running on the edges at both sides of the groove. This idea is visualized in the insets, that render the modal shapes (time averaged electric field amplitude) at 0.6 μm. At this wavelength the first mode is not close to WPP(∞) and the hybridization does not take place, but it is already happening for the second mode. The described phenomenon is even more distinct in Fig. 4.11 displaying the fundamental mode for increasing wavelengths. It is observed that the CPP mode becomes more and more mixed with the WPP(∞). Close to cutoff (at about 1.44 μm) the mode is not guided at the groove bottom anymore but rather at the groove edges. A hint of this possibility was mentioned in the first experimental work [52]. In the experiments, the edges at both sides of the groove have larger radius of curvature than in the previously presented simulations. We have verified that this does not alter our conclusion by repeating the same computation with a radius of curvature of 100 nm at the groove edges (while keeping a radius of curvature of 10 nm at the bottom). Figure 4.11d shows the instantaneous transverse electric field amplitude for this case and it is clear that hybridization with edge modes still occurs. The transverse electric field is approximately horizontal inside the channel (an assumption used by the effective index approximation), but it is not horizontal near the edges where the field is maximum. Let us note in passing the excellent agreement between of the two techniques employed here (the residual discrepancy in Fig. 4.10 for the fundamental mode at 0.6 μm is due to different rounding schemes of the groove bottom in the two methods). From the point of view of fabrication it is useful to mention that, for $\lambda \in (0.6, 0.8\,\mu\text{m})$, the dispersion relation is extremely sensitive to the fine details of the groove bottom (e.g., rounding), as concluded after a large number of simulations where the details of the bottom were subjected to small perturbations. On the other hand, this does not happen for telecom wavelengths (as expected from the modal shape), a circumstance that has also been observed experimentally [29]. Note that the calculated cutoff wavelength of the fundamental mode is somewhat lower than the wavelengths used in the experiments. This discrepancy can be ascribed to (small) differences in the groove geometry, both in the groove shape (angle, side flatness) and in the groove depth, and/or different dielectric permittivity of gold. We have verified (not shown here) that slightly less negative ε or/and smaller groove angle ϕ leads to a higher cutoff wavelength. Finally, the experiments were conducted at ambient conditions so that water condensation could not be excluded (a very thin water layer can significantly increase the cutoff wavelength).

The effect of absorption is summarized in Fig. 4.12 that renders the propagation length $l = [2\text{Im}(k_z)]^{-1}$ versus wavelength, for the various structures considered (k_z is the modal wave vector). The propagation lengths are in all cases smaller than

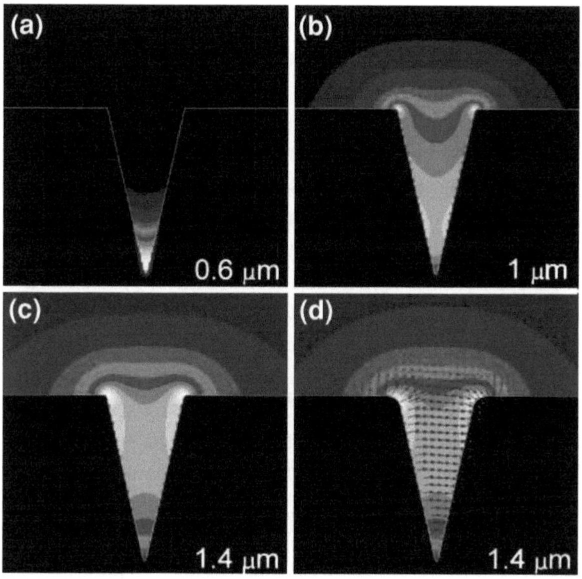

Fig. 4.11 Modal shape of the CPP fundamental mode for increasing wavelength λ. **a** $\lambda = 0.6\,\mu$m, **b** $\lambda = 1\,\mu$m, **c** $\lambda = 1.4\,\mu$m (close to cutoff). These panels display the time averaged electric field amplitude. **d** Instantaneous transverse electric field amplitude at $\lambda = 1.4\,\mu$m for a structure with groove edges rounded with 100 nm radius of curvature. All panels have a lateral size of 2 μm. From Ref. [17]

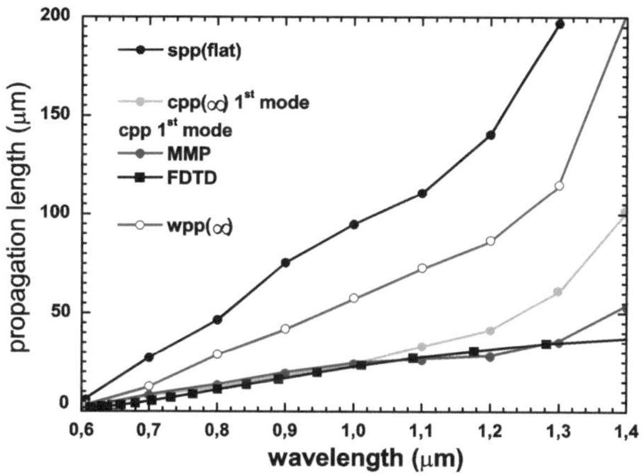

Fig. 4.12 Propagation length versus wavelength for various modes. *Black thick line* SPP mode on a flat surface. *Blue line* (*full circles*) CPP fundamental mode for a groove of height 1.17 μm with the MMP method. *Green line* (*squares*) CPP(∞) fundamental mode for an infinitely deep groove. *Red line* (*open circles*) WPP(∞) mode for an infinitely deep wedge. The FDTD calculation is plotted with *square black* symbols for the truncated groove

that of SPPs at a flat surface. This is a consequence of the field enhancement at the corners and the field confinement that decreases the portion of field propagating in air. When comparing the CPP modes it is observed that the effect of truncation at a finite height is only important for wavelengths larger than 1 μm, which is reasonable because the field is strongly confined at the groove bottom for smaller λ. For longer wavelengths the CPP propagation length is decreased as compared to that of CPP(∞). At $\lambda = 1.4\,\mu$m we find that $l_{\mathrm{CPP}} = 53\,\mu$m and $l_{\mathrm{CPP}} = 35\,\mu$m with MMP and FDTD (See Sect. 1.2.3 for further details), respectively. To explain this discrepancy between the two methods, we must take into account that high l_{CPP} values are a consequence of the "long-lasting" processes being involved in building the resonance up. The different rounding schemas used, while not affecting the dispersion relation spectral position of CPPs at telecom wavelengths, become nevertheless an important issue when propagation lengths are calculated. Note that the experimental values reported in Ref. [28] at 1.55 μm are larger than the computed ones. The discrepancy can be again ascribed to slight differences in geometry and/or dielectric permittivity that rise the cutoff wavelength. If the trend of the line corresponding to the CPP is extrapolated, we find good agreement with the reported data. It must be noticed that the propagation length of WPP(∞) is significantly higher, a fact that could find obvious applications as we will show later.

4.3.2 Wedge Plasmon Polaritons

The basic structure studied here is a metallic wedge surrounded by vacuum. It has an infinitely long edge, which is the propagation direction (Z-axis) for the supported electromagnetic modes (the edge is rounded with radius of curvature r). The wedge angle is denoted as ϕ. We consider both wedges truncated at a certain height $y = h$ (as shown in Fig. 4.8b), and non-truncated wedges ($h \to \infty$). The modes corresponding to non-truncated wedges will be named WPP(∞), as before. The considered metal is gold. The size of the considered structures is sufficiently large so as to use bulk dielectric functions and neglect additional damping due to electron scattering at the metal surface. The effective index model allows one to argue that a metallic wedge sustains modes that are localized close to its edge and propagate along it, but numerical simulations are needed to determine accurately the modal characteristics. The results presented in this section have been obtained with FDTD and MMP.

After an early analysis of WPPs in the electrostatic approximation [49], these modes were studied by Pile *et al.* [23] in the visible regime, where the mode propagation length is very short. Here the emphasis is on telecom wavelengths where losses are much lower (see also [53]). Figure 4.13 displays the modal behavior of WPP(∞) modes for a wedge with angle $\phi = 20°$ and radius of curvature of the edge $r = 10$ nm. Panel (a) shows the dispersion relation (red line) of the fundamental mode. As corresponds to a non-radiative mode, it lies outside the shaded area bounded by the dispersion relation of a SPP mode. The mode has no cutoff wavelength. The modal shape for wavelengths at both ends of the considered spectrum is

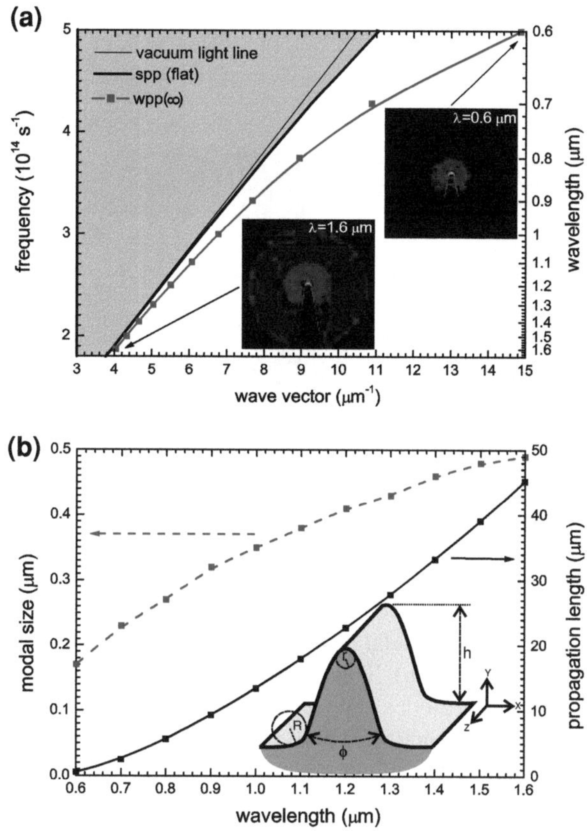

Fig. 4.13 **a** WPP dispersion relation. *Black thick line*, SPP mode on a flat surface; *red line* (*squares*), WPP(∞) mode supported by a non-truncated wedge. *Insets* time-averaged electric field amplitude of WPP(∞) mode at wavelengths $\lambda = 0.6\,\mu$m, and $\lambda = 1.6\,\mu$m. The lateral size of the *insets* is $0.5\,\mu$m. **b** Modal size (*red dashed line*) and propagation length (*black solid line*) of WPP(∞) mode as a function of λ. *Inset* diagram of the truncated wedge

plotted in the insets. Modal size and propagation length as a function of wavelength are presented in panel (b) (left and right axes, respectively). Here the modal size is defined as the transverse separation between the locations where the electric field amplitude of the mode has fallen to one tenth of its maximum value. The factor $1/10$ in this definition is somehow arbitrary but it is sufficient for our mode characterization purposes. The modal size (red dashed line) grows as λ increases, but subwavelength guiding is achieved in the whole regime. As mentioned above, the propagation length (black line) is very short in the visible region of the spectrum ($\simeq 0.5\,\mu$m at $\lambda = 0.6\,\mu$m), but it rises to about $40\,\mu$m in the telecom regime.

Wedge and channel plasmon-polaritons are now compared at $\lambda = 1.5\,\mu$m. In order to have a meaningful comparison, we consider exactly the same geometry for both structures, simply exchanging the metallic and vacuum regions (Fig. 4.14). The angle and radius of curvature are the same as above, but we now compute more realistic structures with finite height ($h = 1.2\,\mu$m, a typical value for experiments with CPPs). The corners where the flat horizontal surface meets the triangular structure are also rounded (with a radius of curvature $R = 100$ nm, also typical for experimental CPPs).

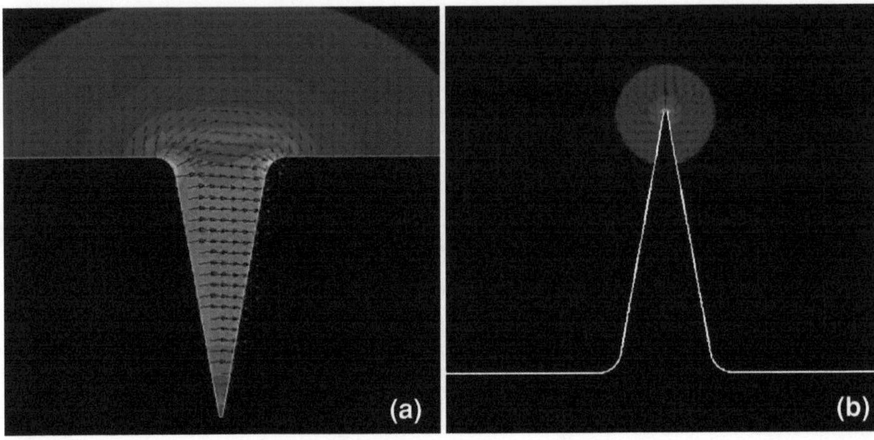

Fig. 4.14 Transverse electric field amplitude of **a** CPP mode, and **b** WPP mode, both at $\lambda = 1.5\,\mu$m. The geometry of both structures is identical (see main text) exchanging the metallic and vacuum regions. The lateral size of panels is 2 μm

The transverse electric field amplitude of both CPP and WPP modes is plotted in Fig. 4.14. The modal size of the WPP is 0.46 μm, significantly smaller than that of the CPP (2.5 μm). This is mainly due to the fact that the CPP mode is hybridized with wedge modes supported by the edges at both sides of the groove, as stated in the previous section. These edges correspond to wedges with a large angle ($\phi' = 100°$) and radius of curvature ($R = 100$ nm), and for which the corresponding WPP modal sizes are larger, as will be shown below. Despite the different modal sizes, the computed propagation lengths are quite similar for both modes: 37 μm for the WPP, and 34 μm for the CPP. It is worth mentioning that the CPP mode is very close to cutoff and for a height $h = 1\,\mu$m the mode is no longer guided, whereas the WPP is guided whenever the height verifies $h > h_c \simeq 0.2\,\mu$m.

Let us now consider the dependence of the modal characteristics of WPPs as a function of the most relevant geometric parameters. The following data correspond to $\lambda = 1.5\,\mu$m. The dependence with the height h of the wedge is summarized in panels (a) and (b) of Fig. 4.15, whereas the dependence with the angle ϕ of a WPP(∞) is presented in the lower panels (c) and (d). The modal effective index n_{eff} (i.e., modal wavevector divided by wavevector in vacuum) is displayed in panel (a). As h decreases, n_{eff} tends to the effective index of a SPP on flat surface (for $h < h_c$, n_{eff} reaches the effective index of a SPP and the mode is no longer guided). Note that a low effective index is equivalent to a more extended field, as confirmed in panel (b) where the modal size is plotted (red dashed line). The propagation length is also shown in panel (b) (black line), increasing when the cutoff height is approached. The behavior of the WPP(∞) modal characteristics as the angle ϕ increases is reminiscent to what occurs when the height h decreases. There is however a major difference: there is no critical angle above which the mode is not longer guided. As ϕ is increased towards 180°, propagation length, n_{eff}, and modal size tend to those of a SPP on a flat surface.

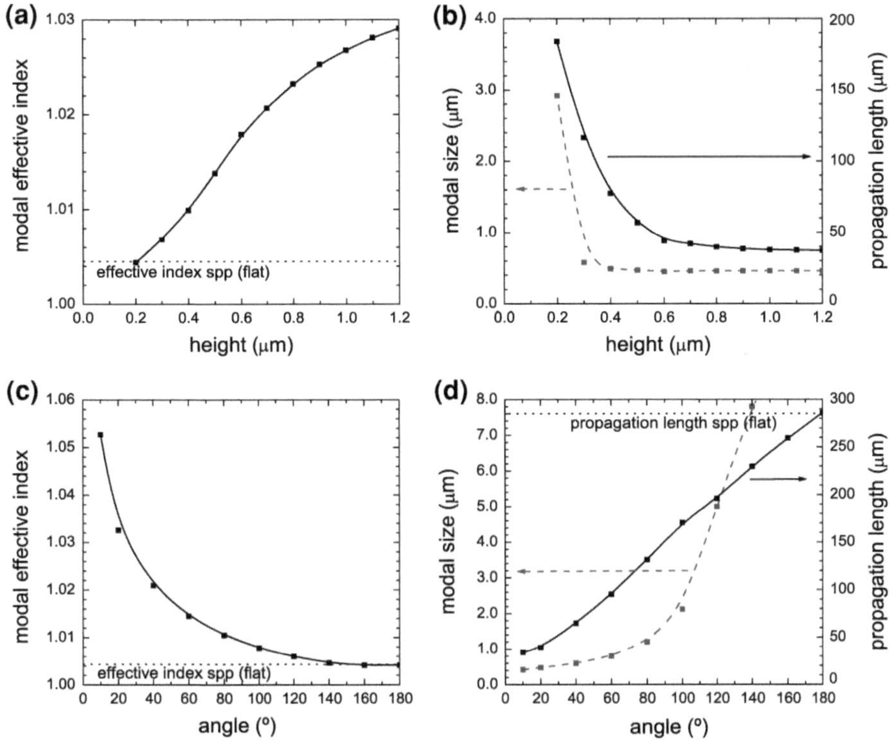

Fig. 4.15 WPP modal characteristics as a function of wedge height or angle. In **a** and **b** the height is varied and the wedge angle is constant, $\phi = 20°$. In **c** and **d** the wedge angle is varied (the wedge is not truncated). The radius of curvature is $r = 10\,\text{nm}$ in all cases. **a** *Solid line*, effective index of WPP mode. **b** *Red dashed line*, modal size of WPP; *solid line*, propagation length of WPP. **c** *Solid line*, effective index of WPP(∞) mode. **d** *Red dashed line*, modal size of WPP(∞); *solid line*, propagation length of WPP(∞)

Modal size rapidly increases as the angle grows, but our numerical simulations show waveguiding no matter how large the angle is (whenever $\phi < 180°$).

4.3.3 CPP and WPP Based Devices

A WPP to SPP Conversion Device

Up to this point we have only been concerned with waveguiding, finding that WPPs display better confinement as compared to CPPs, while keeping both similar propagation lengths. In this section we present strategies for WPP$\leftrightarrow$ SPP conversion. A device with this functionality should convert the shape and size of a WPP into that of a SPP. The task can be also understood as focusing a SPP to a WPP (if the time arrow is reversed). Since SPPs on flat surfaces have infinite transverse extension, the main challenge is the large modal mismatch. Here, conversion will be achieved by deforming the metal surface from a wedge geometry to a flat geometry in a continuous way along the mode propagation direction (*Z*-axis). In other words, the wedge height or angle become functions of the *z*-coordinate, $h(z)$, $\phi(z)$. It is clear from Fig. 4.15a and c that, as the angle ϕ increases or the height h shrinks, the effective index is reduced, leading to a growth of the modal size. If this conversion were done adiabatically, radiation and reflection losses would be negligible, but absorption losses would be large. We hope that a conversion performed within a few wavelengths would strongly reduce absorption losses, while maintaining low reflection and radiation. In order to verify this scenario we have performed three-dimensional (3D) FDTD simulations for structures with constant wedge angle ($\phi = 20°$) and various $h(z)$ profiles. The tight confinement of the mode requires very fine meshes (we used a mesh of 10 nm). The simulation domain is a parallelepiped surrounded by perfect matched layers (PMLs). The simulations are performed at $\lambda = 1.5\,\mu$m in continuous-wave mode. All structures have an initial section with constant wedge height (See Fig. 4.16). The input WPP mode is excited by a source (located at $z = 0.5\,\mu$m in front of the wedge and buried inside a cavity to prevent direct illumination from the source to the conversion device).

After a short spatial transient ($z < z_t = 2\,\mu$m), the field settles down to a propagating WPP mode. We first simulate a set of test structures all of them with constant height, $h \in (0.2, 0.48)\,\mu$m (structures I). These computations will allow the evaluation of losses in later structures, and serve as a test of the FDTD simulations (as the comparison with MMP results, which were done for 2D systems, is very good). After the initial transient, the exponential modal decay found is solely due to absorption in the metal, and not in the PMLs. Modal reflection at the domain boundary is very small (reflected power less than 0.1%). Thus, for structures I, the field computed at $z = 9.0\,\mu$m (shortly before the simulation domain boundary) is a pure WPP mode without radiation. Structure II has $h = 0.48\,\mu$m for $z < 3.9\,\mu$m, and then the wedge height decreases linearly to zero along a distance $\Delta z = 3.3\,\mu$m ($h(z)$ is zero beyond this point). Figure 4.16a shows a longitudinal cross section at the *ZY* plane (side view), while Fig. 4.16b shows the same but at the *ZX* plane at a height $h = 0.3\,\mu$m (top view) of the electric field amplitude. The mode runs from left to right. From panels (a) and (b) reflection seems to be very low. As the wedge height decreases, the modal size expands. This is observed in panels (c–f), which display the corresponding transverse cross sections (*XY* planes at various *z*-coordinates). For comparison, we have considered structure III, which is identical to structure II for $z < 4.6\,\mu$m but with a wedge height abruptly becoming zero after this point (side and top view in panels (g) and (h), respectively). In this case we observe strong reflection at the discontinuity.

We now want to evaluate non-ohmic losses during WPP→SPP conversion in structure II. To this end, for every transverse cross section $z < z_c \simeq 5.8\,\mu$m (z_c being the coordinate corresponding to the modal cutoff height h_c), we have computed the overlap of the field in structure II at that *z*, and the field of a WPP mode with the

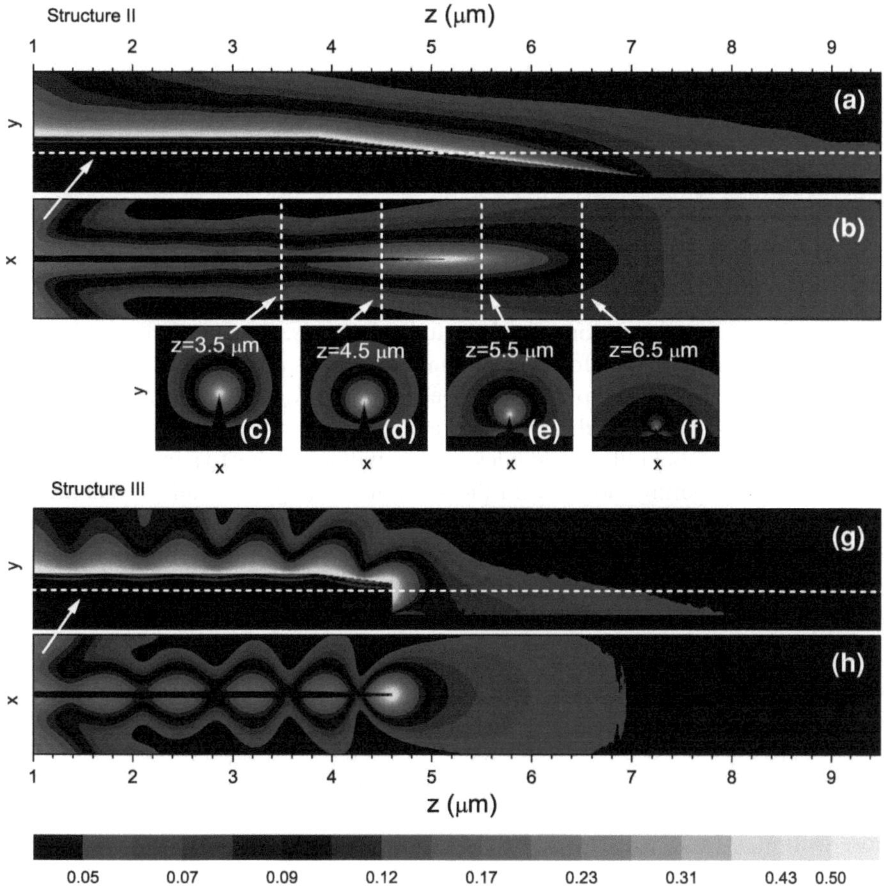

Fig. 4.16 Electric field amplitude in the WPP→ SPP transition (geometries are detailed in the main text). **a** and **b** longitudinal cross sections for structure II (*dashed line* in (**a**) is the location of longitudinal section (**b**)). Transverse cross sections (*XY* plane, at various *z*-coordinates) for structure II are rendered in (**c–f**) (*dashed lines* in (**b**) show the position of these transverse cross sections). Longitudinal cross sections for structure III are rendered in (**g**) and (**h**). The cartesian axes are shown in Fig. 4.8b. The wavelength is $\lambda = 1.5\,\mu$m. The size of all panels along *X* and *Y* directions is $1.4\,\mu$m

height of structure II at z (these fields are available from our computations of type I structures). Let us briefly describe the method used here to calculate field overlaps. It can be shown that the electromagnetic eigenmodes supported by the translational symmetric structure (e.g., a wedge of constant height) are mutually orthogonal [54]. Let us denote such eigenmodes as

$$|n\rangle = |n(\mathbf{r_T})\rangle = \{\mathbf{E}_n(\mathbf{r_T}), \mathbf{H}_n(\mathbf{r_T})\} \tag{4.5}$$

where $n = \pm 1, \pm 2, \pm 3, \ldots$. The fundamental mode is $n = \pm 1$, and negative indices correspond to modes propagating in the negative z direction. $\{\mathbf{E}_n, \mathbf{H}_n\}$ stands for the electric and magnetic field, and $\mathbf{r_T} = (x, y)$ are coordinates in the transverse plane. Eigenmode orthogonality reads

$$\begin{aligned}\langle n|m\rangle &= \langle n(\mathbf{r_T})|m(\mathbf{r}_T)\rangle \\ &= \int\int_{XY\text{plane}} dxdy\mathbf{e_z}\mathbf{E}_n(\mathbf{r_T}) \times \mathbf{H}_m^*(\mathbf{r_T}) \quad = \text{sgn}(n.m)\delta_{|n||m|}\end{aligned} \tag{4.6}$$

where $\mathbf{e_z}$ is a unit vector along the longitudinal Z axis, the star denotes complex conjugate, and sgn(.) stands for the sign function. Let us remark that: (1) the dependence on the z-coordinate, $\exp(ik_n z)$, has been omitted (k_n is the modal wave vector), (2) orthogonality applies both for guided and radiation modes (continuous indices should be used to label radiation modes, but we will avoid this to simplify notation), (3) counter propagating modes with the same index (e.g., $|1\rangle$ and $|-1\rangle$) are not orthogonal, (4) the scalar product of a mode with itself is proportional to the power carried in the longitudinal Z direction, (5) in general, the integral should be carried out in the infinite transverse XY plane. Nevertheless, when one of the modes is guided the integrand is non-negligible only in a finite part of the XY plane, due to transverse localization of the guided mode. Thus, in our computations of scalar products shown later, the integration area will be the transverse FDTD simulation window, (6) in fiber and guided optics, orthogonality conditions are routinely used even when small losses are present. For a general non-cylindrical structure (e.g., a wedge with height varying along the z-coordinate), a generic solution $|f(x, y, z)\rangle$ can be expanded in eigenmodes. For each z the eigenmodes corresponding to that particular transverse cross section, $|n(x, y, z)\rangle$, should be used:

$$|f(x, y, z)\rangle = \sum_n a_n(z)|n(x, y, z)\rangle \tag{4.7}$$

where the coefficients $a_n(z)$ in the linear expansion are related to the projections (also termed overlaps) of the solution $|f\rangle$ on the various eigenmodes $|n\rangle$. For instance, the overlap with the fundamental WPP mode ($n = +1$) is

$$\langle f|1\rangle(z) = \langle f(x, y, z)|1(x, y, z)\rangle = \int\int_{XY\text{plane}} dxdy\mathbf{e_z}\{\mathbf{E_f}(\mathbf{x}, \mathbf{y}, \mathbf{z}) \times \mathbf{H}_1^*(\mathbf{x}, \mathbf{y}, \mathbf{z})\} \tag{4.8}$$

When absorption is present, it is convenient to normalize both $|f\rangle$ and $|1\rangle$ in a particular way that simplifies the bookkeeping of radiation leakage. Namely, at every transverse cross section, $z = const$, the functions $|f\rangle$ and $|1\rangle$ are normalized to unity in the chosen finite integration area. In the following we will plot the square of the overlap integral, $|\langle f|1\rangle(z)|^2$, for the structures considered here. Notice that, since $|1\rangle$ and $|-1\rangle$ are not orthogonal, this function may include an oscillating term whenever reflection occurs, due to the interference of both eigenmodes and the

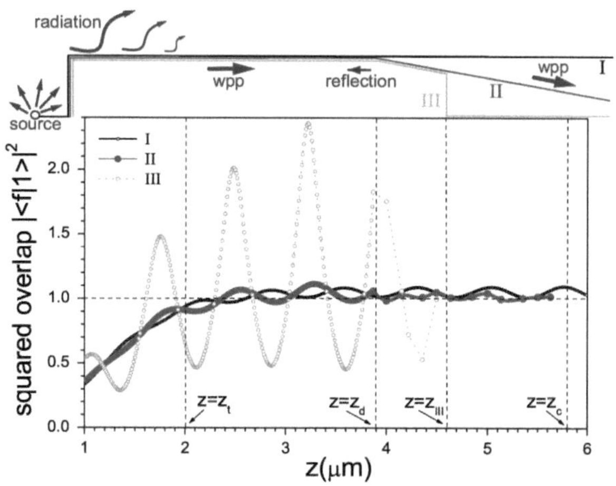

Fig. 4.17 *Squared* overlap integral $|\langle f|1\rangle(z)|^2$ as a function of the longitudinal z-coordinate for various structures. *Black solid line* structure I, *red line* structure II, *green dotted line* structure III. The schematics on *top* of the graph shows the height profile for the three structures considered and the physical processes involved

subsequent formation of a standing wave. On the other hand, the function should be constant for single mode propagation, with no reflection and negligible radiation losses (the mentioned constant is unity with the chosen normalization). This function is also smaller than unity when the linear expansion of $|f\rangle$ includes other modes different from $|\pm 1\rangle$. In other words, $|\langle f|1\rangle(z)|^2$ is smaller than unity when radiation is present (in the chosen finite normalization area). This is expected in regions close to the source. As we move away from it, it is expected the contribution of radiation to the total field diminishes, thereby $|\langle f|1\rangle(z)|^2$ should tend to a unit value.

Figure 4.17 renders the function $|\langle f|1\rangle(z)|^2$ for the three mentioned structures. For the function associated to structure *I* (black line) we distinctly observe three phenomena: (1) small ripples, (2) a value lower than unity for $z < z_t = 2\,\mu$m, and (3) a value about unity for $z > z_t$. The ripples are due to the interference of the incoming WPP and a reflected (counter propagating) WPP at PMLs. The spatial period of the oscillation is consistent with the WPP wave vector. From the amplitude of the ripples it can be computed that the reflection coefficient is 0.1%. The function being smaller than unity for $z < z_t$ is due to the fact that, in our FDTD simulations, the source excites both WPP and radiation modes. The displayed behavior of $|\langle f|1\rangle(z)|^2$ shows that the contribution of radiation modes to the total field $|f(x, y, z)\rangle$ is negligible (in the transverse simulation window) after the excitation transient (i.e., for $z > z_t$). Finally, a value of the function about unity for $z > z_t$ demonstrates that, after the excitation transient, radiation does not leak anymore (ohmic absorption is the only source of losses after the excitation transient in structure *I*). The analysis of structure *II* (red line) is analogous: the reflection is still very small (0.2%) and not important for our purposes. This tiny reflection is most likely caused by the discontinuity (at $z_d = 3.9\,\mu$m) in our conversion device. The radiation losses in the excitation transient are similar to those discussed for structure *I*. Finally, the function $|\langle f|1\rangle(z)|^2$ is plotted as long as the WPP mode exists (i.e., for $z < z_c = 5.8\,\mu$m).

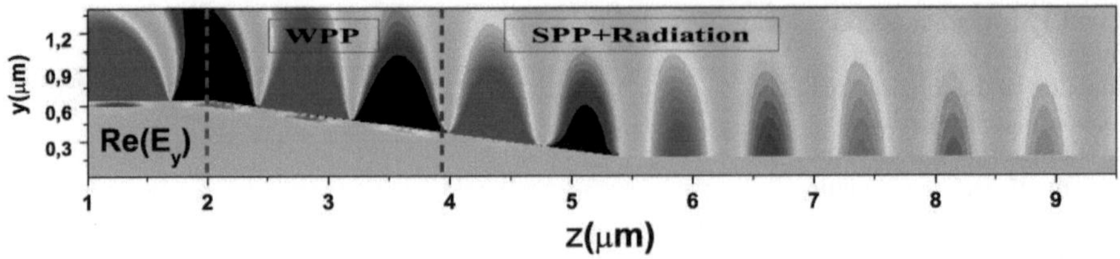

Fig. 4.18 Near field image of the vertical (dominant) component of the electric field

The graph demonstrates that radiation leakage induced by our conversion device is very small: less than 5% for $z_t < z < z_c$. These are the important results of this section: our $WPP \rightarrow SPP$ conversion device produces very little amount of reflection and radiation up to the coordinate z_c, where the WPP mode reaches the cutoff and the field extends outside the simulation window. The data corresponding to structure *III* (green dotted line) show large oscillations due to the reflection of the WPP mode at the abrupt height discontinuity ($z_{III} = 4.6\,\mu$m). Reflection is estimated to be about 20% in this case. In summary, reflection and radiation can be estimated with the help of overlap integrals. The performed tests show that the $WPP \rightarrow SPP$ conversion device with varying height produces very small reflection and radiation losses.

Besides, quantitative comparison between the field computed for structure II in the transverse cross section at $z = 9\,\mu$m and that of a pure SPP shows that the field is mainly a SPP at this coordinate. It is difficult to distinguish between radiation, SPP, and WPP along the remaining length available in the simulation domain, as all these modes have similar wavelengths at the chosen telecom wavelength ($\lambda = 1.5\,\mu$m). As an aid for the visualization Fig. 4.18 shows a cross section of the real part of the electric field dominant component (the vertical one). Qualitatively, the field after the end of the ramp very much resembles a SPP. A quantitative proof that a SPP is excited follows. We have compared the decay along the vertical (Y) direction of the field computed for structure II at various z-coordinates, and the same decay for a pure SPP on a flat surface (we have done the comparison for all vector components for both E and H fields). Figure 4.19 renders the dominant (vertical) component of the electric field along vertical lines located at increasing z-coordinates (colored curves) and the same magnitude for a SPP (computed analytically, black curve). Two comments are in order: (1) for increasing z the decay of the curves resembles more and more an exponential function, which would not be the case if the field was pure radiation. In other words, this suggests that, for increasing z, as radiation escapes, the field is more akin to a SPP, (2) the agreement is not perfect for large Y, which is also expected since interference with a certain amount of radiation is still present (the behavior of the other important field component (H_x) is very similar). The field at the transverse simulation window at $z = 9\,\mu$m is, strictly speaking, a superposition of SPP and radiation, but the main contribution is plasmonic (Fig. 4.19). This information can be used to estimate the power carried by SPPs through this transverse cross section. We have seen that the radiation contribution at the surface ($z = 9\,\mu$m, $y = 0$) is

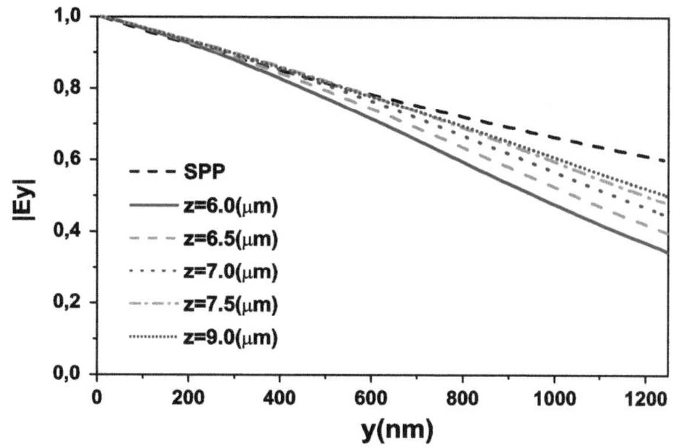

Fig. 4.19 Dominant (vertical) component of the electric field along *vertical lines* located at increasing z-coordinates (*colored curves*) and the same magnitude for a SPP (computed analytically, *black curve*)

negligible and there the field is essentially a SPP. Assuming that the field at the surface ($z = 9\,\mu$m, $y = 0$, obtained from the simulation) corresponds to a pure SPP mode, the power carried by this SPP can be evaluated as 29 % of the input power. This value fits very well with that obtained in the simulation for the transmitted power.

This agreement is consistent with the similarity of the fields shown in Figure 4.19, (power scales with the square of the field). We can thus conclude that the power through the simulation window at $z = 9\,\mu$m is essentially carried by SPPs. Notice that, at the end of the device, an (in-plane) angular spectrum of SPPs is excited, so that the total power coupled as SPPs may be larger than 29%. The finite size of the simulation domain impedes the calculation of the total power coupled to SPPs propagating in all directions in the horizontal plane. In any case, the goal of the device is converting the WPP mode into a SPP propagating collinear with the Z-axis.

Finally, in structure II, WPP modes do not exist for $h < h_c$. Moreover, such a device converts a WPP mode to SPPs propagating with an in-plane angular spectrum. For both reasons the coupling of WPP to Z-propagating SPP may be reduced. These restrictions should not apply for WPP(∞) modes when the angle is the control parameter along the Z-axis. Therefore, we expect a better performance for a structure where the wedge angle ϕ is continuously varied from a flat surface ($\phi = 180°$) to a wedge with $\phi = 20°$. Three-dimensional simulations of this case are inherently very difficult because, by construction, the modal size grows increasingly fast as $\phi \rightarrow 180°$. Thus, the mode cannot fit in the simulation domain, which is constrained by the available computer memory. Nevertheless, the idea is illustrated in Fig. 4.20, that plots the transverse electric field for decreasing wedge angles (2D MMP simulations). Note how the field is concentrated close to the edge. The previously shown FDTD computations and other studies [55], lead us to expect that focusing (without radiation or reflection) can be achieved in a short length also in this case.

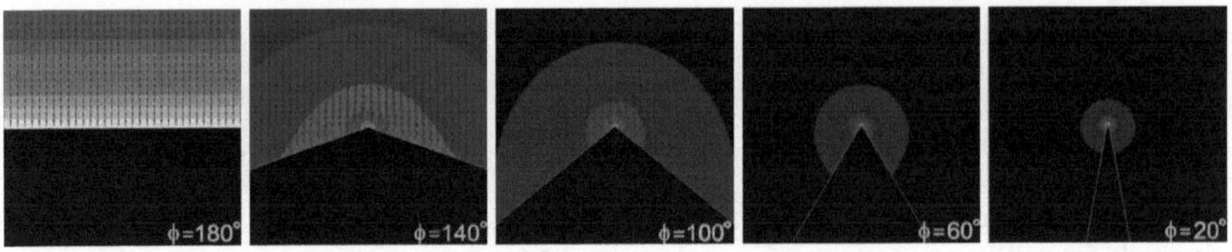

Fig. 4.20 Focusing of the WPP(∞) modal field as the wedge angle ϕ is slowly decreased from $\phi = 180°$ (SPP mode on a flat surface) to $\phi = 20°$. The panels display the transverse electric field at a wavelength of 1.5 μm. The lateral size of the panels is 2 μm. The radius of curvature of the tip is $r = 10$ nm. The color scale of the various panels is not the same

The Plasmonic "Candle"

Nanoguiding and concentrating optical radiation with SPP modes supported by metal nanostructures is a main strategic research direction in Plasmonics [3, 56], with implications ranging from quantum optics [57] to nano-sensing [58]. Various configurations have been suggested for SPP nanofocusing [18, 59–63] all of them supporting progressively more confined SPP modes in the limit of infinitely small waveguide cross sections. However, experimental demonstrations of SPP nanofocusing [64, 65] have so far been indirect (based on far-field observations of scattered [64] or frequency upconverted [65] radiation) and inconclusive with respect to the field enhancement achieved in the focus. Here we report a mechanism for radiation nanofocusing with CPPs that propagate along subwavelength metal grooves being gradually tapered synchronously both in depth and width. Efficient CPP nanofocusing at telecom wavelengths is directly demonstrated.

The idea of radiation nanofocusing (and thereby of greatly enhanced electromagnetic fields) by gradually decreasing a waveguide cross section has always been very appealing due to its apparent simplicity. Its realization however requires for the corresponding waveguide mode to scale in size along with the waveguide cross section, a nontrivial characteristic that is not readily accessible and, for example, cannot be achieved with dielectric waveguides due to the diffraction limit. The physics of SPP guiding is fundamentally different and intimately connected with the hybrid nature of SPP modes, in which electromagnetic fields in dielectrics are coupled to free electron oscillations in metals [66]. Several SPP guiding configurations exhibit, in the limit of infinitely small waveguide dimensions, the required scale invariance, i.e., the mode size scaling linearly with that of the waveguide. The appropriate SPP modes are supported, for example, by thin metal films (short-range SPPs) and narrow gaps between metal surfaces (gap SPPs) [67], and by corresponding cylindrical, i.e., rod and coaxial, structures [20, 68]. Note that their nanofocusing [59–63] is conceptually simple only at a fairly basic level and requires dealing with several rather complicated issues, such as excitation of the proper SPP mode [64] and balancing between SPP propagation losses (that increase for smaller waveguide cross sections) and focusing effects [69]. The situation becomes even more complicated if one considers SPP modes whose scaling behavior is not straightforward. Thus CPP guides, which

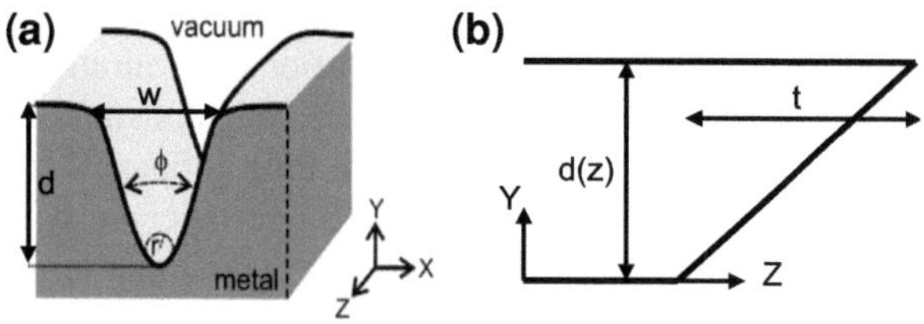

Fig. 4.21 **a** Schematic of the V-groove geometry and **b** the taper region

can be efficiently excited with optical fibres and used for ultracompact plasmonic components [29], exhibit rather complicated behavior with respect to their geometrical parameters, as we have seen in Sect. 4.3.1, and their potential for nanofocusing of radiation has not yet been explored. An important quantity in waveguide theory is the so-called "waveguide parameter", *V*, which is a measure of the field confinement in a particular waveguide. This parameter was previously used for planar thin-film waveguides as a basic parameter that, along with the asymmetry parameter, allowed charting universal dispersion curves for TE modes [70]. In this case the normalized waveguide parameter is in the form: $V = k_0 w \sqrt{\epsilon_f - \epsilon_s}$, w being the film thickness, ϵ_f and ϵ_s the dielectric constants of the film and the substrate, respectively. Based on the last expression, we approach the problem of CPP nanofocusing by making use of the following (approximate) expression for the normalized (CPP-based) waveguide parameter [71]:

$$V_{\text{CPP}} \cong 2\sqrt{\frac{k_0 d \epsilon_d \sqrt{|\epsilon_d - \epsilon_m|}}{|\epsilon_m| tan(\theta/2)}} \equiv 4d\sqrt{\frac{\pi \epsilon_d \sqrt{|\epsilon_d - \epsilon_m|}}{\lambda w |\epsilon_m|}} \quad (4.9)$$

λ is the light wavelength ($k_0 = 2\pi/\lambda$), d and w are the V-groove depth and width, θ is the groove angle so that $tan(\theta/2) = 0.5w/d$, ϵ_d and ϵ_d are the dielectric constants of dielectric and metal (See Fig. 4.21a). It has been demonstrated that V-grooves with different dimensions and operating at different wavelengths but having the same waveguide parameter feature very similar field confinement [71]. If the CPP waveguide groove depth and angle are gradually and synchronously decreased, the corresponding waveguide parameter (Eq. 4.9) could be kept constant. In this case (and within the same approximation [71]), one can show that the CPP effective index of nanometer-sized V-grooves diverges towards the taper end: $n_{\text{CPP}} \approx (k_0 d)^{-1}$, i.e., it behaves in the same manner as that of the SPP mode of a tapered nanowire [60]. In the adiabatic approximation, the CPP field is continuously squeezed by the walls of a tapered V-groove with the maximum field being limited only by the CPP propagation loss. In general and similarly to the nanofocusing with nanowires [69], the field enhancement at the taper end is a result of the interplay between CPP dissipation (contributed to by CPP absorption, reflection and out-of-plane scattering) and field squeezing.

In order to gain further insight and reveal the potential of CPP nanofocusing, we have conducted 3D-FDTD simulations for V-grooves terminated with tapers of different lengths. In the considered configuration, the metal (gold) surface is deforming from the straight channel geometry to a flat surface in a continuous way along the mode propagation direction, i.e., along the Z-axis (Fig. 4.21a). In other words, the channel parameters are kept constant ($d_0 = 1\,\mu\text{m}, \phi_0 = 28°, w_0 \cong 450\,\text{nm}$) during the initial 20- μm-long propagation and then become functions of the z-coordinate. We considered linear tapering with respect to the groove depth: $d(z) = d_0(1 - (z - z_0)/t)$, where t is the taper length, z_0 is the starting coordinate of the taper and $z_0 < z < z_0 + t$ (Fig. 4.21b). At the same time, the groove width was adjusted following four different dependencies:

$$w(z) = w_0(d(z)/d_0)^n \tag{4.10}$$

for $n=0, 1, 2$ and 3. Consequently, the normalized waveguide parameter introduced in Eq. 4.9 varied as follows: $V_{\text{CPP}}(z) = V^0_{\text{CPP}}(d_0/d(z))^{0.5n-1}$. Therefore, as the channel tip is approached, the parameter V was maintained constant for $n = 2$, while $V \mapsto 0$ for $n = 1$ (constant-angle tapering) and $n = 0$ (constant-width tapering), finally $V \mapsto \infty$ for $n = 3$.

The tight confinement of the mode requires very fine meshes (we used a mesh of 10 nm), for a working wavelength in the telecom regime. The simulation domain is a parallelepiped surrounded by PMLs to avoid spurious reflections on the system boundaries. The fundamental CPP mode is excited by a monochromatic oscillating magnetic dipole source pointing along the z direction and located at $z = 1\,\mu$m into the channel. Such a light source has the same E-field symmetry of a CPP, so it efficiently couples to CPPs. The excitation wavelength is chosen to be 1,480 nm.

Near field information can be readily retrieved once the stationary state is reached, as it is shown in Fig. 4.22. In this figure we can see field distributions through the lateral (y–z plane) cross sections where the groove width and depth decrease in accordance with the different dependencies described with (a) $n = 0$, (b) $n = 1$, (c) $n = 2$, (d) $n = 3$. We have chosen the 2-μm-long taper because it features the best enhancement in all cases, as we will see later on. Besides the occurrence of a very bright spot in near-field (Fig. 4.22), these simulations show the increase of the standing wave pattern contrast for the taper with $n = 2$ (that is obviously related to a corresponding increase of the reflection efficiency of the taper) as well as a strong scattering in the taper region (especially for $n = 0$ and 1). However, the simulated V-groove taper (with optimum parameters: $t = 2, n = 2$) characteristics are much better with respect to the field enhancement.

Normalized to the non-tapered channel CPP optical signals ($|E|^2$) for different n values are shown in Fig. 4.23a. These optical signals are extracted from Fig. 4.22 along the dashed line drawn in panel (a) as reference, which is placed at 10 nm above the metal surface. As we can see, the best performance (with respect to the field enhancement achieved at the taper end) is found for $n = 2$, obtaining a maximum value of ~1,000. This result is reinforced by the dependence of the optical signal at its maximum value through the direction normal to the surface, which is depicted in Fig. 4.23b. In this figure is clearly seen the evanescent character of fields above the surface for $n = 2$ and 3. In contrast, for the cases $n = 0$ and $n = 1$ the field near the surface does not decay exponentially because radiation dominates

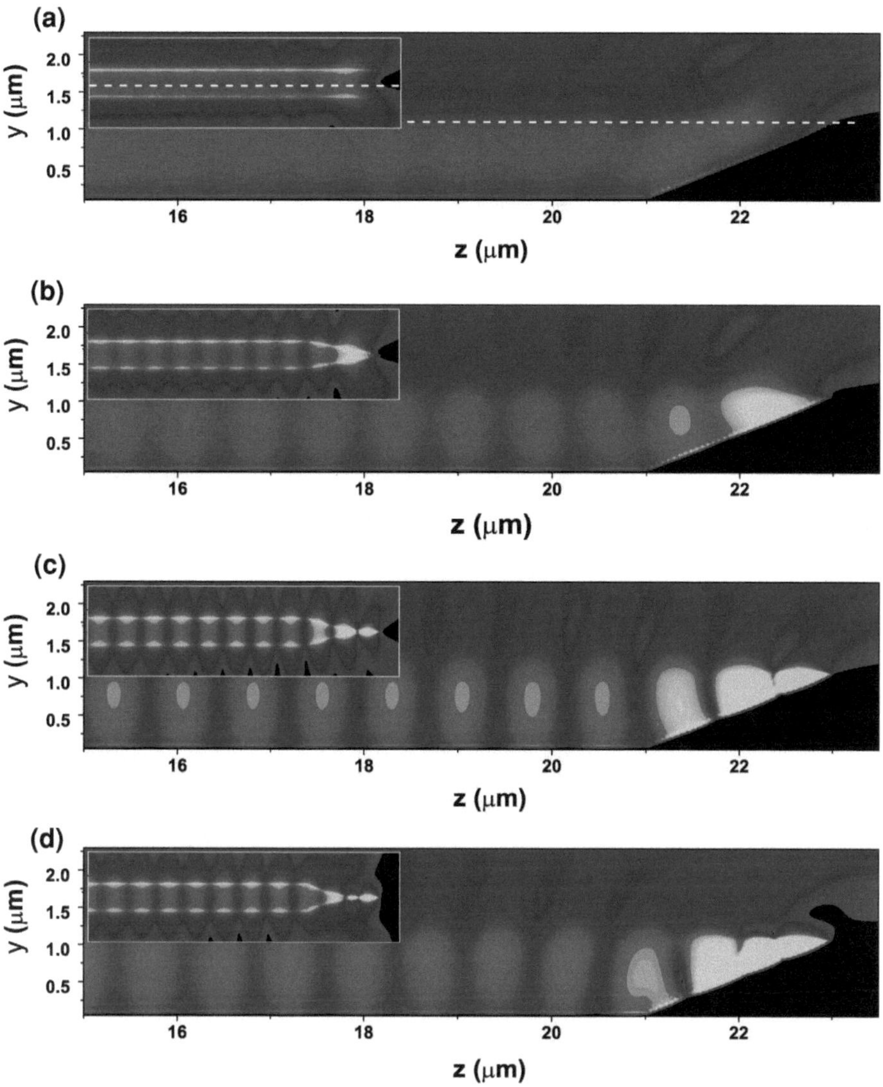

Fig. 4.22 3D-FDTD simulations of CPP field distributions through the lateral (y–z plane) cross sections calculated for 2-μm-long tapers (λ = 1,480 nm). The groove width and depth decrease in accordance with the different dependencies described by Eq. 4.10 with **a** $n = 0$, **b** $n = 1$, **c** $n = 2$, **d** $n = 3$. *Insets* show the corresponding near-field maps at sample surface plane ~10 nm over the surface (indicated with *dashed line* in the main figure **a**)

over reflection at the taper end. The field near the surface does not exponentially decay because of radiation dominates. The theoretical findings agree well with our qualitative considerations in the sense that the groove tapering should be conducted

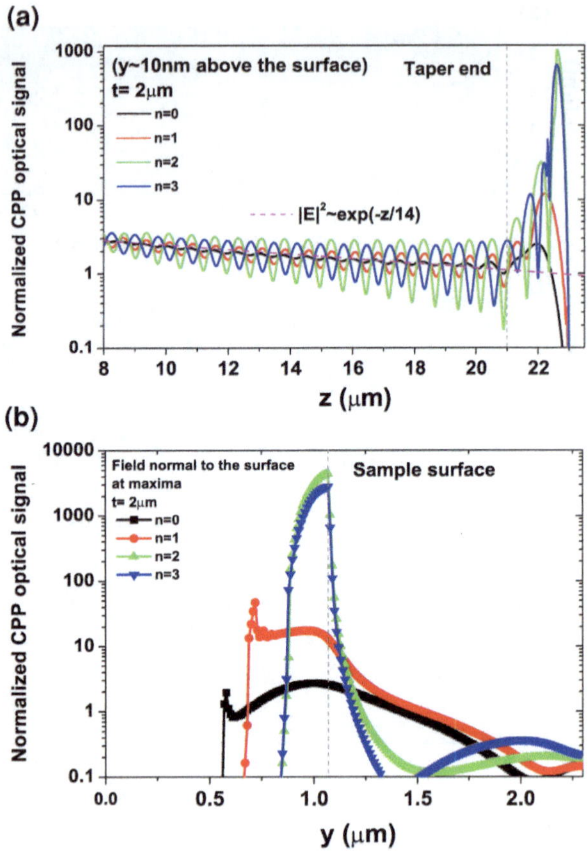

Fig. 4.23 **a** Depicts the normalized CPP optical signal at 10 nm distance over the flat metal surface ($y \sim 1\ \mu$m) along the propagation direction (z), for the 2-μm-long taper and different n values. The *dashed line* shows the exponential fitting of the optical signal along the non-tapered channel. **b** Optical signal through the direction perpendicular to the surface (y) at maxima (**a**)

so that the groove depth and width decrease in accord, keeping the normalized CPP parameter (Eq. 4.9) constant.

Concerning the optimum taper length, we believe that its value is mainly controlled by the constructive interference of the propagating (towards the taper) and reflected CPP modes, though it is affected by the CPP propagation loss as well. Figure 4.24 shows different cross sections of the field amplitude ($|E|$) along the propagation direction at 10 nm above the metal surface ($y\sim1\ \mu$m) for different taper lengths, for the case $n = 2$ (Note that the beginning of the ramp is in this case placed at $z = 1\ \mu$m). As we advanced, the greatest enhancement is achieved for the 2-μm-long taper.

For the experimental verification of the idea, different samples were fabricated using focused ion-beam (FIB) milling by the group of Prof. T.W. Ebbesen at the Laboratoire de Nanostructures, ISIS (Universitè Louis Pasteur). The samples were done in a 1.8-μm-thick gold layer, deposited on a glass substrate coated with ITO. Several straight 150-μm-long V-grooves were milled with the angles close to 28° and depths of 1.1–1.3 μm, which were gradually tapered out over different distances $t = 2, 3, 4$, and 6 μm. The fabricated structures were characterized with a

Fig. 4.24 Cross sections of the field amplitude along the z direction at 10 nm above the metal surface for different taper lengths, being $n = 2$

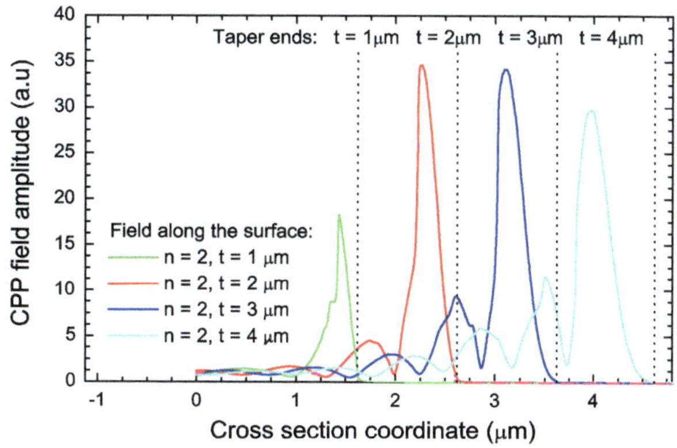

collection scanning near-field optical microscope (SNOM) by the group of Prof. S.I. Bozhevolnyi in Aalborg (Denmark) at the Department of Physics and Nanotechnology. Experimental details on the SNOM setup can be found elsewhere [29].

The SNOM investigations showed that all fabricated structures exhibited the effect of signal enhancement at the taper end, with the near-field optical images featuring subwavelength-sized bright spots located at the taper end as judged from the (simultaneously recorded) topographical images (Fig. 4.25a and b). The largest enhancement of the signal was observed for the and 2-, 3-μm-long tapers (Fig. 4.25c and d) in accordance with FDTD results. The CPP propagation length was estimated to be $\approx$50 μm. This values was obtained from the exponential fit to the optical signal variation along the CPP propagation, for different tapered V-grooves excited at the wavelength of 1,480 nm. While this value is consistent with previous observations [29], it does not agree with the theoretical one appearing in Fig. 4.23, which is about 14 μm. This discrepancy may be due to the theoretical structure having a more acute angle which, as we have investigated in Sect. 4.3.1, implies higher absorption levels. The experimental value is also consistent with the measured signal enhancement Γ, defined as the ratio between the maximum signal in the tapered V-groove and that expected at the taper end coordinate in the absence of tapering (Fig. 4.25e).

In any case, we should emphasize that the relationship between near-field optical signal distributions and field intensity distributions existing near the sample surface (in the absence of a SNOM probe) is very complicated [72], as it is not obvious how the EM fields couple to the SNOM tip. Even in a very simple approximation of the dipole-like detection (taking place at the position of an effective detection point inside a fibre probe) [73], different field components contribute differently to the detected signal, making it impossible to directly relate the near-field intensity distributions and the corresponding SNOM images. In this situation, it is extremely important to control that the detected signal does originate from the evanescent field components (and thereby is associated with the focused CPP fields), since the detection of propagating waves, such as scattered at the taper, is much more efficient than that of evanescent ones [73]. Near-field optical images were recorded with shear force feedback, a few

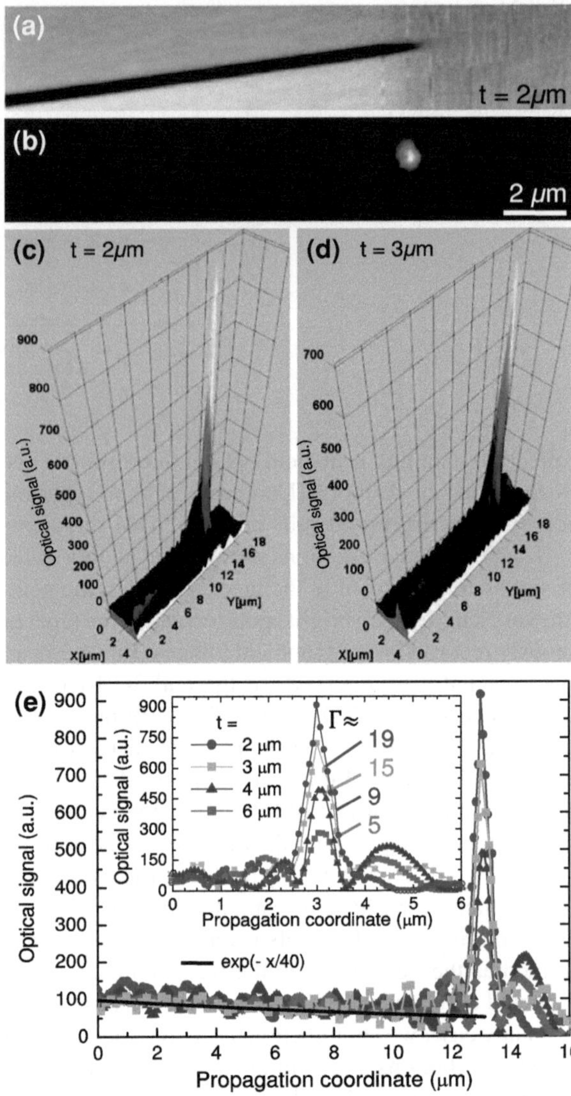

Fig. 4.25 Plasmonic candle. **a** Topographical and **b, c** Near-field optical ($\lambda = 1{,}480$ nm) SNOM images in different presentations in order to emphasize the signal enhancement Γ realized at the taper end. **d** The optical SNOM image obtained with the 3-μm-long taper. **e** Renders the normalized cross sections obtained with the optical images (similar to those shown in (**c**) and (**d**)) recorded for the tapered V-grooves, with different taper lengths

nanometers away from the surface, and then with the SNOM fibre probe scanning along a plane located $\approx$100 nm from the sample surface (Fig. 4.26).

A drastic signal decrease, and significant image blurring, observed with the increase of the probe-surface distance signifies unambiguously that the bright spots seen on the SNOM images are indeed the result of detection of evanescent (CPP) field components. It is further seen that, for the 2-μm-long taper, the maximum optical signal (at the bright spot) decreased by a factor of $\approx$6 (cf. Fig. 4.26b and c) while the CPP-related signal measured away from the taper region decreased only

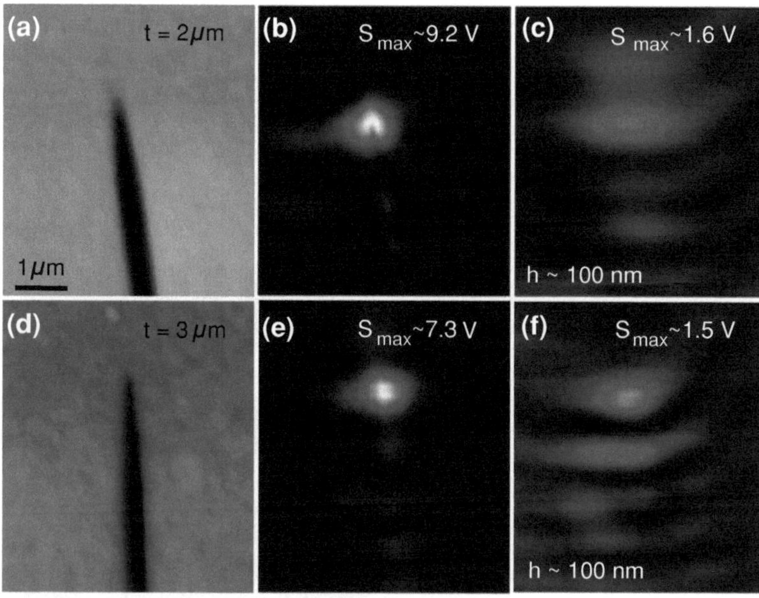

Fig. 4.26 Influence of evanescent field components at telecom ($\lambda = 1{,}480$ nm). **a** and **d** Topographical images for the 2- and 3-μm-long taper samples, respectively. Their corresponding nearfield optical SNOM images (with shear force feedback) are shown in (**b**) and (**e**). **c** and **f** Same as in (**b**) and (**e**) but at 100-nm distance from the sample surface with the tapered V-grooves. The decrease in signal and the significant image blurring when increasing the probe-surface distance reveal the dominance of evanescent field components in the images obtained

by a factor of $\approx$2. It is reasonable to assume that, in both cases, the optical signals (being proportional to the field intensity at an effective detection point [73]) decrease exponentially with the probe-surface distance but at different rates, because the corresponding optical fields are laterally confined to the different widths. The latter implies that the observed intensity enhancement decreases also exponentially with the height of observation plane. Finally, taking into account the circumstance that the effective detection point is located typically $\approx$150 nm away from the tip end [73], we obtained $\approx$90 as a ballpark estimate of the field intensity enhancement realized at the sample surface with the 2-μm-long taper. It is worth to mention that the simulation results are consistent with the experimental observations, which feature (1) the strongest enhancement for the 2-μm-long taper, (2) the occurrence of a very bright spot in near-field optical images and (3) the interference fringes indicating the CPP reflection and scattering in the taper region (cf. Fig. 4.26 and Fig. 4.22c). At the same time, the field enhancement estimated from the calculations is much larger than even the ballpark estimate. Indeed, using the same definition as before one obtains (Fig. 4.23a) the computed field intensity enhancement of $\approx$ 1,000 for the 2-μm-long taper with $n = 2$. Interestingly the maximum values is reached inside the channel, as Fig. 4.23b shows. On the other hand, some difference should be expected given the limited FIB

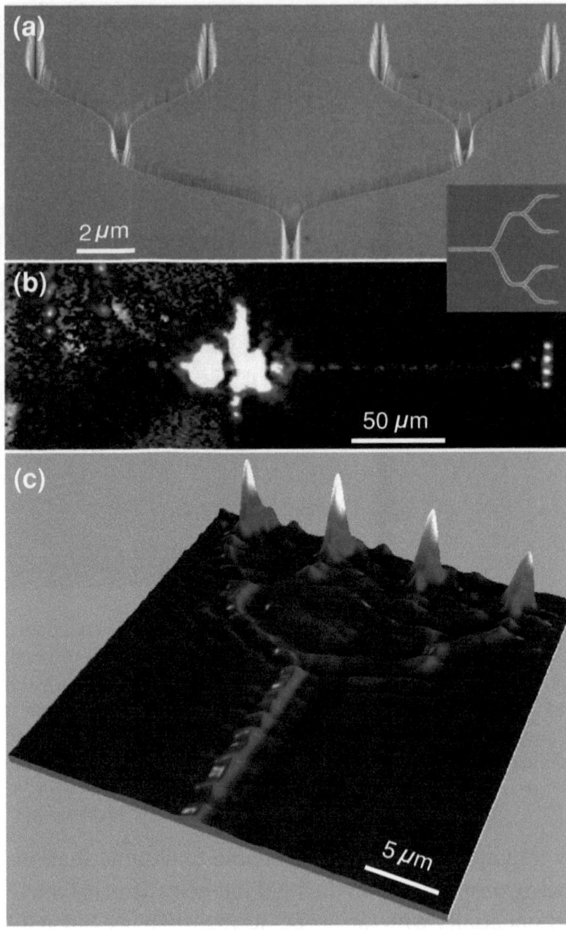

Fig. 4.27 Plasmonic candlestick. **a** SEM image of a multichannel configuration for delivering nanofocused and enhanced radiation to four spatial locations via consecutive 5- μm-long Y-splitters terminated with 2-μm-long tapers (see the *inset* with an overview SEM image). **b** Microscope image of a coupling arrangement superimposed with the far-field image taken at the excitation wavelength $\lambda = 1{,}500$ nm with an infrared camera, showing the track of CPP propagation and four bright spots at the tapers. **c** Near-field optical ($\lambda = 1{,}500$ nm) SNOM image demonstrating significant signal enhancements realized at the four groove tapers, with signal levels being similar and exceeding greatly even the signal level at the input channel

resolution and the fact that the maximum field intensity is calculated to be fairly close to the taper end with the taper width being only $\approx$50 nm (while the groove depth is still $\approx$300 nm).

Finally, it have been explored the prospect of realization of a multichannel configuration for delivering nanofocused and enhanced CPP fields to several different spatial locations by making use of consecutive *Y-splitters* (Fig. 4.27). The level of signal enhancement observed with the SNOM images was fairly constant for the four tapers amounting to a factor $\approx$5 with respect to the signal at the input channel, which is consistent with the enhancement of $\approx$20 observed for the individual 2-μm-long taper (Fig. 4.25c), given the power distribution between four channels. This experiment demonstrates that the suggested approach for radiation nanofocusing is rather versatile and robust, features that are extremely important for future applications.

4.4 Conclusions

First, we have proposed an efficient unidirectional source for SPPs. The way to achieve this is coupling-in with SPPs by means of a subwavelengt aperture in the back-side illumination montage. This proposal allows the modulation of such coupling-in by means of a finite array of grooves. Our approach is based on a simple wave interference model that, despise of the simplified description of some of the physics involved, has been found in good agreement with both sophisticated computer simulations and experimental measurements at NIR and telecom ranges.

Additionally, we have investigated the guiding properties of the plasmonic modes supported by straight metallic grooves and wedges. We have briefly discussed the results obtained by two different rigorous numerical techniques. The dependence on various geometric parameters of the modal dispersion, size, polarization, and losses have been analyzed. The knowledge of these properties can be of great help for the analysis of plasmonic devices featuring edges and corners. Both CPP and WPP modes are quite well confined in the transverse plane and their losses are reasonably low. These properties make them very interesting candidates for plasmonic interconnects.

Finally, we have also explored the possibility of light focusing via the geometry-driven conversion of a standard SPP into a tightly confined WPP. On the other hand, based on CPP supporting structures we have demonstrated that intense EM fields can be achieved by tapering a channel in a proper manner. This approach has been found rather versatile and robust as it would be desired for future applications. Indeed, one can envisage further development of these concepts for other plasmonic waveguides based on gap SPP modes [71] as well as applications for miniature bio-sensors.

References

1. W. Barnes, A. Dereux, T. Ebbesen, Nature **424**, 824 (2003)
2. E. Ozbay, Science **311**, 189 (2006)
3. T.W. Ebbesen, C. Genet, and S.I. Bozhevolnyi, Phys. Today **61**, 44–50 (2008)
4. F. Lopez-Tejeira, S.G. Rodrigo, L. Martin-Moreno, F.J. Garcia-Vidal, E. Devaux, T.W. Ebbesen, J.R. Krenn, I.P. Radko, S.I. Bozhevolnyi, M.U. Gonzalez et al., Nature Phys. **3**, 324 (2007)
5. F. Lopez-Tejeira, S.G. Rodrigo, L. Martin-Moreno, F.J. Garcia-Vidal, E. Devaux, J. Dintinger, T.W. Ebbesen, J.R. Krenn, I.P. Radko, S.I. Bozhevolnyi et al., New J. Phys. **10**, 033035 (2008)
6. A. Otto, Z. Phys **216**, 398 (1968)
7. B. Lamprecht, J.R. Krenn, G. Schider, H. Ditlbacher, M. Salerno, N. Felidj, A. Leitner, F.R. Aussenegg, J.C. Weeber, Appl. Phys. Lett. **79**, 51 (2001)
8. R.H. Ritchie, E.T. Arakawa, J.J. Cowan, R.N. Hamm, Phys. Rev. Lett. **21**, 1530 (1968)
9. H. Ditlbacher, J.R. Krenn, N. Felidj, B. Lamprecht, G. Schider, M. Salerno, A. Leitner, F.R. Aussenegg, Appl. Phys. Lett. **80**, 404 (2002)
10. C. Sönnichsen, A.C. Duch, G. Steininger, M. Koch, von G. Plessen, J. Feldmann, Appl. Phys. Lett. **76**, 140 (2000)
11. E. Devaux, T.W. Ebbesen, J.-C. Weeber, A. Dereux, Appl. Phys. Lett. **83**, 4936 (2003)
12. L. Yin, V.K. Vlasko-Vlasov, A. Rydh, J. Pearson, U. Welp, S.-H. Chang, S.K. Gray, G.C. Schatz, D.B. Brown, C.W. Kimball, Appl. Phys. Lett. **85**, 467 (2004)

13. E. Popov, N. Bonod, M. Nevière, H. Rigneault, P.-F. Lenne, P. Chaumet, Appl. Opt. **44**, 2332 (2005)
14. A. Agrawal, H. Cao, A. Nahata, New J. Phys. **7**, 249 (2005)
15. S.-H. Chang, S. Gray, G. Schatz, Opt. Express **13**, 3150 (2005)
16. P. Lalanne, J.P. Hugonin, J.C. Rodier, Phys. Rev. Lett. **95**, 263902 (2005)
17. E. Moreno, F.J. Garcia-Vidal, S.G. Rodrigo, L. Martin-Moreno, S.I. Bozhevolnyi, Opt. Lett. **31**, 3447 (2006)
18. E. Moreno, S.G. Rodrigo, S.I. Bozhevolnyi, L. Martín-Moreno, F.J. García-Vidal, Phys. Rev. Lett. **100**, 023901 (2008)
19. V.S. Volkov, S.I. Bozhevolnyi, S.G. Rodrigo, L. Martín-Moreno, F.J. García-Vidal, E. Devaux, T.W. Ebbesen, Nano Lett. **9**, 1278 (2009)
20. J. Takahara, S. Yamagishi, H. Taki, A. Morimoto, T. Kobayashi, Opt. Lett. **22**, 475 (1997)
21. P. Berini, Opt. Lett. **24**, 1011 (1999)
22. D.F.P. Pile, D.K. Gramotnev, Opt. Lett. **29**, 1069 (2004)
23. D.F.P. Pile, T. Ogawa, D.K. Gramotnev, T. Okamoto, M. Haraguchi, M. Fukui, S. Matsuo, Appl. Phys. Lett. **87**, 061106 (2005)
24. D.F.P. Pile, T. Ogawa, D.K. Gramotnev, Y. Matsuzaki, K.C. Vernon, K. Yamaguchi, T. Okamoto, M. Haraguchi, M. Fukui, Appl. Phys. Lett. **87**, 261114 (2005)
25. A. Manjavacas, de F.J.G. Abajo, Nano Lett. **9**, 1285 (2009)
26. I.V. Novikov, A.A. Maradudin, Phys. Rev. B **66**, 035403 (2002)
27. D.F.P. Pile, D.K. Gramotnev, Opt. Lett. **30**, 1186 (2005)
28. S.I. Bozhevolnyi, V.S. Volkov, E. Devaux, T.W. Ebbesen, Phys. Rev. Lett. **95**, 046802 (2005)
29. S.I. Bozhevolnyi, V.S. Volkov, E. Devaux, J.Y. Laluet, T.W. Ebbesen, Nature **440**, 508 (2006)
30. A. Boltasseva, V.S. Volkov, R.B. Nielsen, E. Moreno, S.G. Rodrigo, S.I. Bozhevolnyi, Opt. Express **16**, 5252 (2008)
31. C. Hafner, *Post-Modern Electromagnetics* (Wiley, Chichester, 1999)
32. S. Maier, Curr. Nanosci. **1**, 17 (2005)
33. J.-C. Weeber, J.R. Krenn, A. Dereux, B. Lamprecht, Y. Lacroute, J.P. Goudonnet, Phys. Rev. B **64**, 045411 (2001)
34. J. Krenn, H. Ditlbacher, G. Schider, A. Hohenau, A. Leitner, F. Aussenegg, J. Microsc. **209**, 167 (2003)
35. J.-C. Weeber, Y. Lacroute, A. Dereux, E. Devaux, T. Ebbesen, C. Girard, M.U. González, A.-L. Baudrion, Phys. Rev. B **70**, 235406 (2004)
36. J.G. Rivas, M. Kuttge, H. Kurz, P.H. Bolivar, J.A. Sánchez-Gil, Appl. Phys. Lett. **88**, 082106 (2006)
37. M.U. González, J.-C. Weeber, A.-L. Baudrion, A. Dereux, A.L. Stepanov, J.R. Krenn, E. Devaux, T.W. Ebbesen, Phys. Rev. B **73**, 155416 (2006)
38. L. Aigouy, P. Lalanne, J.P. Hugonin, G. Julié, V. Mathet, M. Mortier, Phys. Rev. Lett. **98**, 153902 (2007)
39. F. López-Tejeira, F.J. García-Vidal, L. Martín-Moreno, Phys. Rev. B (R) **72**, 161405 (2005)
40. F. López-Tejeira, F. García-Vidal, L. Martín-Moreno, Appl. Phys. A Mater. Sci. Process. **89**, 251 (2007)
41. S.C. Kitson, W.L. Barnes, J.R. Sambles, Phys. Rev. Lett. **77**, 2670 (1996)
42. P. Lalanne, J.P. Hugonin, Nat. Phys. **2**, 551 (2006)
43. A.Y. Nikitin, S.G. Rodrigo, F.J. García-Vidal, L. Martín-Moreno, New J. Phys. **11**, 123020 (2009)
44. W. Nomura, M. Ohtsu, T. Yatsui, Appl. Phys. Lett. **86**, 181108 (2005)
45. L. Yin, V. Vlasko-Vlasov, J. Pearson, J. Hiller, J. Hua, U. Welp, D. Brown, C. Kimball, Nano Lett. **5**, 1399 (2005)
46. Z. Liu, J. Steele, W. Srituravanich, Y. Pikus, C. Sun, X. Zhang, Nano Lett. **5**, 1726 (2005)
47. H. Offerhaus, B. vandenBergen, M. Escalante, F. Segerink, J. Korterik, N. vanHulst, Nano Lett. **5**, 2144 (2005)
48. J.M. Steele, Z. Liu, Y. Wang, X. Zhang, Opt. Express **14**, 5664 (2006)

49. L. Dobrzynski, A.A. Maradudin, Phys. Rev. B **6**, 3810 (1972)
50. S.I. Bozhevolnyi, Opt. Express **14**, 9467 (2006)
51. D.K. Gramotnev, D.F.P. Pile, Appl. Phys. Lett. **85**, 6323 (2004)
52. V.S. Volkov, S.I. Bozhevolnyi, E. Devaux, T.W. Ebbesen, Opt. Express **14**, 4494 (2006)
53. M. Yan, M. J. Qiu, Opt. Soc. Am. B **24**, 2333 (2007)
54. A.W. Snyder, J.D. Love, *Optical Waveguide Theory* (Chapman and Hall, London, 1983)
55. S.A. Maier, S.R. Andrews, L. Martín-Moreno, F.J. García-Vidal, Phys. Rev. Lett. **97**, 176805 (2006)
56. S. Lal, S. Link, N.J. Halas, Nat. Photon. **1**, 641 (2007)
57. D.E. Chang, A.S. Sorensen, E.A. Demler, M.D. Lukin, Nat. Phys. **3**, 807 (2007)
58. M. Ringler, A. Schwemer, M. Wunderlich, A. Nichtl, K. Kürzinger, T.A. Klar, J. Feldmann, Phys. Rev. Lett. **100**, 203002 (2008)
59. K.V. Nerkararyan, Phys. Lett. A **237**, 103 (1997)
60. A.J. Babadjanyan, N.L. Margaryan, K.V. Nerkararyan, J. Appl. Phys. **87**, 3785 (2000)
61. M.I. Stockman, Phys. Rev. Lett. **93**, 137404 (2004)
62. D.K. Gramotnev, J. Appl. Phys. **98**, 104302 (2005)
63. K.C. Vernon, D.K. Gramotnev, D.F.P. Pile, J. Appl. Phys. **101**, 104312 (2007)
64. C. Ropers, C. Neacsu, T. Elsaesser, M. Albrecht, M. Raschke, C. Lienau, Nano Lett. **7**, 2784 (2007)
65. E. Verhagen, A. Polman, L.K. Kuipers, Opt. Express **16**, 45 (2008)
66. H. Raether, *Surface Plasmons on Smooth and Rough Surfaces and on Gratings* (Springer, Berlin, 1988)
67. E.N. Economou, Phys. Rev. **182**, 539 (1969)
68. C.A. Pfeiffer, E.N. Economou, K.L. Ngai, Phys. Rev. B **10**, 3038 (1974)
69. D.K. Gramotnev, M.W. Vogel, M.I. Stockman, J. Appl. Phys. **104**, 034311 (2008)
70. H. Kogelnik, V. Ramaswamy, Appl. Opt. **13**, 1857 (1974)
71. S.I. Bozhevolnyi, J. Jung, Opt. Express **16**, 2676 (2008)
72. S.I. Bozhevolnyi, B. Vohnsen, E.A. Bozhevolnaya, Opt. Commun. **172**, 171 (1999)
73. I.P. Radko, S.I. Bozhevolnyi, N. Gregersen, Appl. Opt. **45**, 4054 (2006)

Chapter 5
Optical Field Enhancement on Arrays of Gold Nano-Particles

5.1 Introduction

Light scattering by arrays of metal nanoparticles gives rise to nanostructured optical fields exhibiting strong and spatially localized (on a nanometer scale) field intensity enhancements that play a major role in various surface phenomena. The local field enhancement effects are of high interest, in general, for fundamental optics and electrodynamics [1], and for various applied research areas, such as surface enhanced Raman spectroscopy [2] and microscopy including optical characterization of individual molecules [3]. Furthermore, the highly concentrated EM fields around metallic nanoparticles are thought to enhance, in turn, non-linear effects, which could pave the way for active plasmonic-based technologies. Also biotechnology can take advantage of such high intensified optical fields as for instance, it has been demonstrated in trapping living cells [4]. Individual metal particles can exhibit optical resonances associated with resonant collective electron oscillations known as localized surface plasmons (LSPs) [5]. Excitation of LSPs results in the occurrence of pronounced bands in extinction and reflection spectra and in local field enhancement effects. On the other hand, random arrangements of gold particles on gold film are well known substrates for the observation of strong surface enhanced effects like surface enhanced Raman scattering (SERS) [2, 6] or surface enhanced luminescence [7, 8]. The physical mechanism responsible for the surface enhanced effects is at least partly related to the strong optical near fields close to resonantly driven surface plasmons on nano-structured metal (usually gold or silver) surfaces [5]. On random substrates (e.g. deposited colloid, electrochemically roughened or evaporated films) so-called "hot-spots" are observed, i.e strongly localized areas on the substrate, where the enhancement is particularly strong [6]. For these hot-spots record enhancement values for SERS of up to 10^{14} have been reported [2]. These huge enhancements are considered to be related to narrow gaps between neighboring nanoparticles, which are electromagnetically coupled across the narrow gap thereby causing huge fields inside the gap [6]. This interpretation is qualitatively supported by different simulations, but a detailed understanding of the effect is still missing.

S. G. Rodrigo, *Optical Properties of Nanostructured Metallic Systems*,
Springer Theses, DOI: 10.1007/978-3-642-23085-1_5,

On one hand, numerical simulations are challenging due to computational limits causing restrictions to the minimal cell size and due to the limited knowledge of the (non)local behavior of the metal dielectric function in these size regimes. On the other hand, these very narrow gaps are also experimentally demanding, both in the exact metrological characterization below the 10 nm range and the reproducible fabrication of such structures with nm precision or better. These drawbacks might be addressed by arranging the particles. In this way, the enhancement effects present in random particle distributions on gold may be further enhanced due to resonant interactions between particles periodically arranged. For metal particles placed on a metal surface, inter-particle interactions can be mediated by surface plasmon polaritons (SPPs), whose resonant excitation can be achieved by tuning the array periodicity.

The search for configurations ensuring reliable realizations of strongly enhanced local fields is often conducted with the help of linear extinction/reflection spectroscopy, where minima in the transmitted/reflected light intensity are associated with the excitation of system resonances (see [9] and references therein). This characterization technique is considered reliable but lacking spatial resolution, since the spectra are influenced by a whole illuminated area containing many nanoparticles. In addition, the information obtained is not direct, since the relation between the extinction/reflection minima and local field enhancements is rather complicated. For example, the extinction is determined not only by the absorption and scattering of individual particles but also by the scattering diagram of a given particle array [10]. A more direct approach for the evaluation of local field intensity enhancement has been recently developed [11]. This technique is based on the fact that strongly enhanced local fields due to the excitation of LSPs in gold nano-structures give rise to two-photon absorption. This, in turn, leads to a broad emission continuum generated by inter-band transitions of *d*-band electrons into the conduction band known as two-photon luminescence (TPL) [12–15]. It has been demonstrated that nonlinear scanning optical microscopy, in which the TPL excited with a strongly focused laser beam is detected, can be used for characterization of the local field intensity enhancement in gold nanoparticles (bow-tie nano antennas) and at their surfaces [11]. However, it was difficult to ascertain the accuracy of a main formula used in the developed approach, since a crucial parameter, viz., the area of TPL origin, had to be found in the course of simulations [11]. It would be advantageous to verify this approach by using the same modeling tool for the (same) scattering system investigated not only with the TPL but also with another, preferably well-established experimental technique (e.g., extinction spectroscopy).

In this chapter, we summarize our results [16–18] on the optical response (extinction spectroscopy and TPL emission) of regular arrays of rectangular gold nanoparticles deposited either on glass (Sect. 5.3) or on gold (Sect. 5.4) substrates, which are investigated from the theoretical and experimental point of view. In order to model the optical response, we use the finite-difference-time-domain (FDTD) approach (see Sect. 1.2). The nanoparticle samples were made by Dr. A. Hohenau and Prof. J.R. Kreen at the Karl-Franzens University and Erwin Schrödinger Institute for Nanoscale Research in Graz (Austria). They also characterized the samples by using linear extinction spectroscopy. The TPL measurements were carried out by Dr. J. Beermann and Prof. S.I. Bozhevolnyi from the Department of Physics and Nanotechnology at Aalborg University.

5.2 Sample Description and Methods

5.2.1 Simulations

The structures consist either on gold particles on top of an semi-infinitely extending glass substrate (gold-on-glass) or on gold particles on top of a 55 nm thick gold film deposited also on a glass substrate (gold-on-gold). The fields were calculated in a box of $\Lambda_x \times \Lambda_y \times 1.3\,\mu$m with periodic boundary conditions at the walls perpendicular to the substrate and an "uniaxial perfect matched layer" (UPML) at the walls parallel to the substrate. Additionally, the Complementary Concurrent Operators Method (CCOM) layers were also added to the UPML layers (Sect. 1.2.5). The parameters defining the CCOM layers were chosen complementary to the UPML layers. In this way, one can use thinner UPML layers and still absorb better the energy flowing at grazing incidence. This is of special importance in the considered system, as small SPP peaks in reflection have to be resolved and small errors in the reflection of grazing modes due to unwanted lack of absorption by the absorbing layers could be attributed spuriously to SPP resonances. Typically the cell size was $5 \times 5 \times 5$ nm in space and the time steps were 0.0077 fs (corresponding to a Courant-Friedrich-Levy factor of 0.8), to guarantee the numerical stability of the (3D) simulations. Moreover, we found that after a simulation time of 100 fs, convergency was guaranteed. As usual, the dielectric function for gold was approximated by a Drude-Lorentz formula (see Sect. 1.2.4). In the simulations, the structures were excited by a Gaussian wave-packet composed of plane waves with wave-vector **k** perpendicular to the substrate and all frequencies of interest (Sect. 1.2.2).

Spectra were calculated after projection onto diffracted modes (Sect. 1.2.3). In the comparison with experimental data, only the zero order mode was considered in the post-processing, as experimental intensities were collected in a small solid angle centered around the normal direction.

For the calculation of the TPL enhancement, we assume that the TPL signal originates from the "top" layer (~half a skin depth deep) of the covered gold areas following the surface topography. The achievable TPL intensity enhancement factor $\alpha(\lambda)$ can be evaluated from the electric field amplitudes E by

$$|\alpha|^2(\lambda) = \frac{\int\int_{A_{\text{cell}}} |E(x, y, z_b, \lambda)|^4 dxdy}{\int\int_{A_{\text{cell}}} |E(z_m, \lambda)|^4 dxdy}, \tag{5.1}$$

where the quantity $|E(x, y, z_b, \lambda)|^4$ integrated over the top layer of the gold surface (bump or unit cell) is assumed to be proportional to the TPL-signal from the array

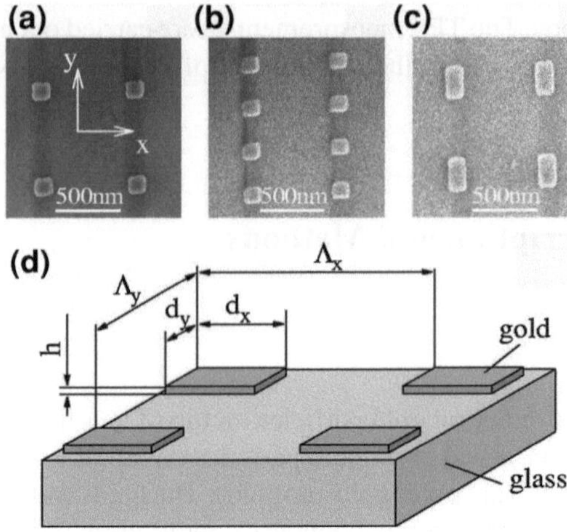

Fig. 5.1 SEM images of the investigated arrays of rectangular gold nanoparticles on glass substrate with **a** $\Lambda_x = \Lambda_y = 740$ nm and $d_x = d_y = 130$ nm (type A), **b** $\Lambda_x = 2\Lambda_y = 740$ nm and $d_x = d_y = 130$ nm (type B) and **c** with $\Lambda_x = \Lambda_y = 740$ nm and $d_x = 0.5d_y = 130$ nm (type C). **d** Sketch of the arrays defining the different parameters. The array and particles dimensions were varied between the different samples in the range of (all dimensions in nm): $130 \leq d_x \leq 160,\ 65 \leq d_y \leq 320\ h = 25$ or $50,\ 740 \leq \Lambda_x \leq 860$ and $370 \leq \Lambda_y \leq 1740$

and compared to $|E(z_m, \lambda)|^4$ integrated over the same size of area, but from a smooth gold film without particles. Using simulated intensity enhancement maps as those will be shown later, but obtained for several excitation wavelengths, it is possible to estimate the spectral dependence of the achievable TPL enhancement.

5.2.2 Experimental

The gold-on-glass particle samples consist of 2D-arrays of nominally rectangular gold particles on top of a 0.5 mm thick glass-substrate (Fig. 5.1) produced by electron beam lithography [19]. The dimensions of nanoparticles ($\sim$150 $\times$ 150 $\times$ 50 nm^3) were chosen to realize the LSP resonance with a polarization parallel to the substrate close to the wavelength of 750 nm, so that its influence could be observed both, in optical spectra and TPL images. On the other hand, the gold-on-gold particle samples consist of 2D-arrays of nominally rectangular, 50 nm high gold particles on top of a 55 nm thick gold film on glass-substrate (*inset* Fig. 5.11b) produced also by electron beam lithography. The overall size of the arrays is 100 $\times$ 100 μm^2. The lateral dimensions of the particles and the grating constants of the arrays were varied to systematically study their influence on the optical extinction and reflection spectra and the TPL signal.

Extinction spectra were recorded by a Zeiss MMS-1 micro-spectrometer attached to a conventional optical microscope equipped with a 2.5×, 0.075 numerical aperture objective. To control the polarization of the incident light, a polarizer is inserted in the optical path of the microscope. The extinction is calculated from the transmission of the arrays on the substrate (T) with reference to the transmission of the bare substrate (T_{glass}) as $\log_{10}\left(T_{\text{glass}}/T\right)$, for the gold-on-glass samples. Importantly the reference for the reflection spectra was taken on the plain, unstructured gold film outside the areas covered with particle arrays, for the gold-on-gold samples.

The experimental setup for TPL scanning microscopy enables to simultaneously record the TPL signal (detected by a photomultiplier) and the backscattered light (detected by a photodiode) as a function of the sample position with the resolution determined by a focal spot size of the excitation laser at the sample surface ($\sim$1 μm) [20, 21]. In these experiments, it was used a 200 fs Ti:Sapphire laser at a repetition rate of 80 MHz with a linewidth of $\sim$10 nm and an adjustable polarization plane. The laser wavelength can be tuned between 720–900 nm, which permits to make spectrally resolved studies of TPL efficiency. The typical average incident power is in the range of 0.1–50 mW.

5.3 Spectroscopy and TPL of Au Nanoparticle Arrays on Glass

In this section, we combine the results of TPL microscopy (in the wavelength range of 720–800 nm) with theoretical modeling based on the FDTD, where only the independently determined geometry and dielectric functions of the sample were used as input without any adjustable parameters. Additionally, the theoretical results are compared carefully to experimentally recorded extinction spectra (in the wavelength range of 450–950 nm) to verify the quality of the modeling. In general, the individual LSP resonances of nanoparticles can be tuned by varying both, particle dimensions and shapes [22], while the resonances stemming from multiple interactions within the grating of particles (scatterers) can be tuned by varying the grating period or gap between particles [9, 11, 23]. We pay special attention to the case when the individual LSP resonance is matched to the grating resonance since one could expect a strong modification of the field intensity enhancement compared to the "normal" case [1, 11].

5.3.1 Spectroscopy

We first investigate numerical extinction spectra compared to the experimental results of the fabricated arrays of gold particles on glass substrate, aiming at identification of the LSP resonances and understanding of their interrelations with the particle shapes and sizes as well as the array periods in analogy to the previous studies of arrays of ellipsoidal particles [5, 10].

To demonstrate the influence of particle shape and array parameters on the extinction spectra, we focus in the following on three different types of arrays (see Fig. 5.1): (A) arrays with $\Lambda_x = \Lambda_y$ and $d_x = d_y$ (quadratic array of quadratic particles), (B) arrays with $\Lambda_x = 2\Lambda_y$ and $d_x = d_y$ (rectangular array of quadratic particles), and (C) arrays with $\Lambda_x = \Lambda_y$ and $2d_x = d_y$ (quadratic array of rectangular particles). For all arrays, we kept $\Lambda_x = 740$ nm and recorded spectra for $d_x = 130, 140, 150$ and 160 nm.

Figure 5.2 depicts the experimental and simulated extinction spectra of arrays of type A. The dominating feature in the experimental spectra is one broad extinction peak with a maximum between 710 and 810 nm, depending on the polarization and particle size. In analogy to ellipsoidal particles, this extinction peak can be attributed to the excitation of the LSP resonance with dominating dipolar character. Since the particles are rectangular in shape, we can assume that several excited LSP-eigenmodes contribute to this extinction peak [5]. With increasing lateral particle dimensions (the particle height was kept constant) we observe a red-shift and an increase in extinction strength, as it would be the case for arrays of simple ellipsoidal particles of increasing size as well. The slight blue shift of the spectra recorded with polarization parallel to the y-axis compared to those recorded with polarization parallel to the x-axis stems from a slight asymmetry in the actual x and y-dimension of the particles.

The simulated spectra (Fig. 5.2b) agree very well, both quantitatively and qualitatively, with the experimental spectra, thought exhibiting small (but distinct) additional dips at 740 nm and at 550 nm. These dips coincide with the excitation of the first grating orders propagating (nearly) parallel to the sample interface on the air and substrate sides of the particles, respectively.

In the case of type A arrays, due to the large grating constant, optical near-field interactions between the particles only weakly influence the spectral position of the LSP resonances [24]. This was verified by comparing the recorded extinction spectra of type A arrays with grating constants ranging from 740 to 860 nm (not shown). Nevertheless, a considerable effect of the grating constant is observed for arrays of type B, whose extinction spectra exhibit not only the expected red shift for increasing particle sizes but also a remarkable difference between the spectra obtained for x- and y-polarizations (see Fig. 5.3). Such a strong effect can be explained as follows.

For type B arrays, the grating constant parallel to the y-axis Λ_y is only 370 nm, i.e., the array periodicity is lower than the wavelength of light in air for (vacuum-) wavelengths larger than 370 nm and of light in the substrate (refractive index $n = 1.52$) for (vacuum-) wavelength larger than 550 nm. Therefore, the coherently excited (dipolar) LSPs can emit in the y-z-plane only to the $0th$ grating order. In addition, for the polarization parallel to the x-axis, diffraction into the first grating order in the x-z-plane is weak due to the dipolar far-field emission characteristic. Therefore, the radiation damping for x-polarization is considerably reduced, resulting in narrower widths and stronger maxima observed for the corresponding extinction peaks (compare Figs. 5.2 and 5.3). At the same time, for y-polarization, the field scattered by one particle is of comparable strength and approximately in counter phase (due to retardation and the phase shift between LSP and exciting field) with the incident field at the position

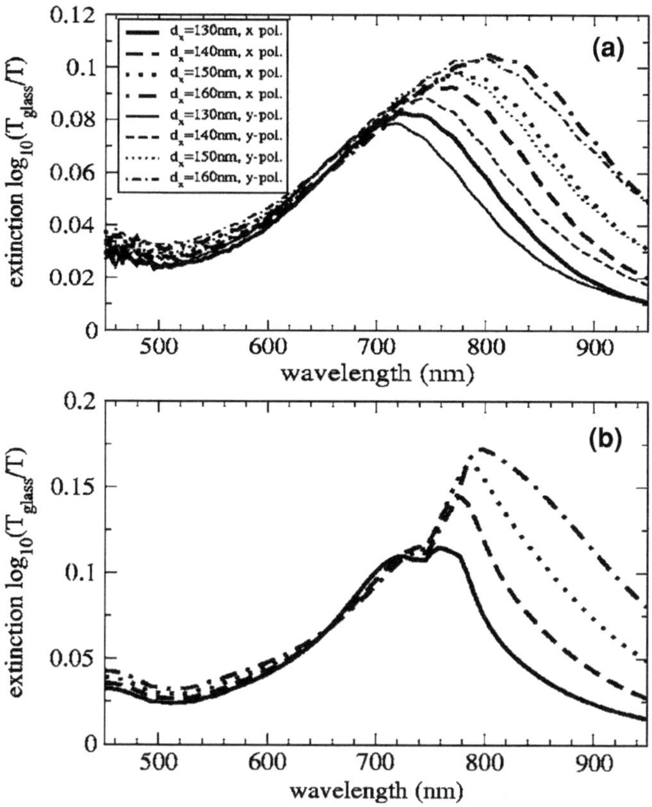

Fig. 5.2 Experimental (**a**) and simulated (**b**) extinction spectra of type A arrays with $\Lambda_x = \Lambda_y =$ 740 nm and $d_x = d_y =$ 130, 140, 150 and 160 nm

of its next neighbors in y-direction, thereby decreasing the total field experienced by the particles. This leads to a decrease of the extinction (normalized to the particle density) compared to arrays of type A.

Also in case of type B arrays we find good agreement of the simulated extinction spectra (Fig. 5.3b) with the experimental ones, except for the behavior observed at 740 nm, where the simulations show strong dips or peaks related to the occurrence of a grazing grating order on the air side. For y-polarization, we find a similar dip in the spectra as for type A arrays, but for x-polarization we observe a sharp peak in case the extinction peak of the particles coincides with the grating constant ($d = 160$ nm; thick, dash-dotted curve).

This behavior can be qualitatively accounted for by the aforementioned effect of coherent superposition of the LSP fields (driven by the incident field) with the fields scattered by the neighboring particles. For the array with the particle size $d = 160$ nm, the particle separation in terms of light wavelength λ at the LSP extinction peak are $\Lambda_y = \lambda/2$ and $\Lambda_x = \lambda$, respectively. Assuming the LSP scattered

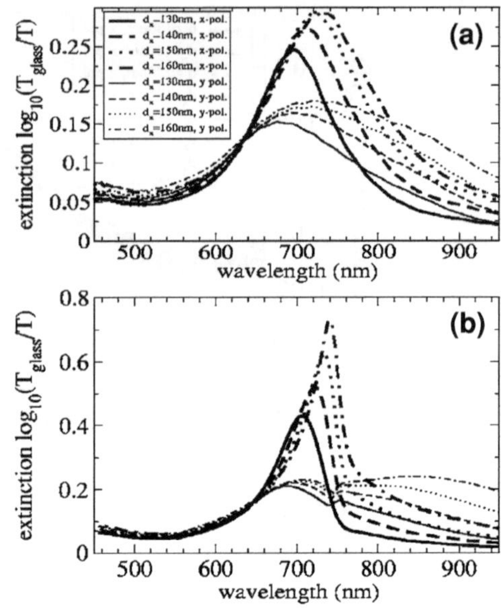

Fig. 5.3 Experimental (**a**) and simulated (**b**) extinction spectra of type B arrays with $\Lambda_x = 2\Lambda_y = 740$ nm and $d_x = d_y = 130, 140, 150$ and 160 nm

fields to be qualitatively similar to the fields emitted by a point dipole, the electric field in the direction of the induced LSP is dominating over that in transverse direction (in this distance regime). Due to retardation, the scattered field of one particle at the position of its next neighbor (in the direction of the dipole) has a phase shift close to π for y-polarization or 2π for x-polarization, respectively. Therefore either destructive or constructive interference occurs which is responsible for the dip or peak in the extinction spectrum. This effect is clearly seen in the distributions of optical near-field intensity calculated for different polarizations. For x-polarization (Fig. 5.4), there is constructive interference of the LSP fields of one particle and the fields scattered by its neighboring particles, similar to the resonance of a series resonant circuit. Therefore the field intensities at the top corners of the particle are very strong. In contrast, for y-polarization (Fig. 5.5), we find (due to destructive interference) very weak field intensities at the top corners of the particle, similar to the antiresonance of a parallel resonant circuit. This spectrally sharp constructive or destructive interference of the calculated local fields also changes the field strength within the particles and is therefore recognized in the far-field by a decrease or increase of the extinction. However, this effect is not readily observed in the experimentally recorded spectra. We will explain this along with other observed discrepancies between experiments and theory in Sect. 5.5.

Finally, let us consider the influence of the particle shape on the spectra. Figure 5.6 displays the experimental and simulated extinction spectra for type C arrays with $\Lambda_x = \Lambda_y = 740$ nm and rectangular particles with $d_y = 2d_x$. Whereas for x-polarization the extinction looks similar to that of type A arrays but with somewhat broader peaks, for y-polarization (parallel to the largest particle dimension) the

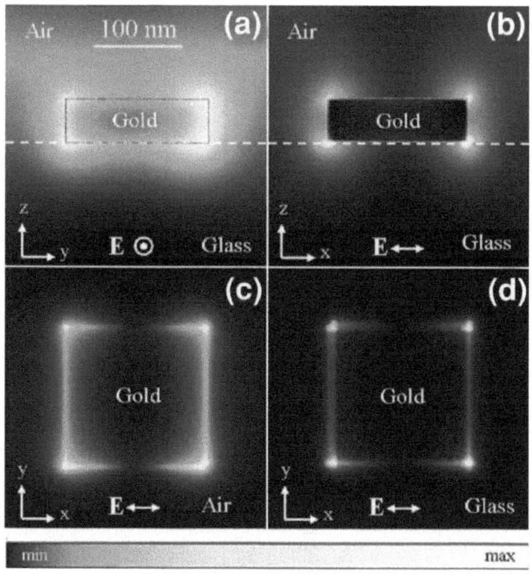

Fig. 5.4 Simulated distribution of the local field intensity enhancement $\frac{|E_{\mathrm{bump}}(\mathbf{r},\lambda)|^2}{|E_{\mathrm{film}}(\lambda)|^2}$ around a single gold particle ($d_x = d_y = 160$ nm, height $d_z = 50$ nm) positioned on a glass substrate in a type B lattice with $\Lambda_x = 2\Lambda_y = 740$ nm and obtained for a x-polarized (resonant case) exciting electric field E as indicated on each image. The distributions are taken in planes either through the *center* of the particle (**a, b**) or from the *top* (**c**) and *bottom* (**d**) surface of the particle as indicated by coordinate axes on each image. The maximum levels are (**a**) ~102, (**b**) ~820, (**c**) ~2300, and (**d**) ~20000

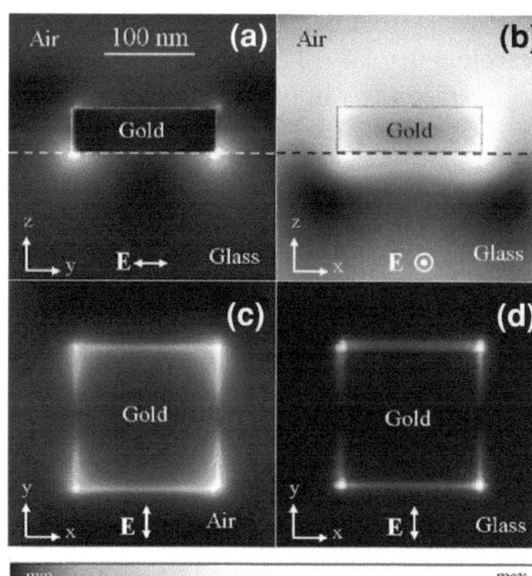

Fig. 5.5 The same as Fig. 5.4, but obtained for a y-polarized (antiresonant case) exciting electric field E as indicated on each image. The maximum levels are **a** ~138, **b** ~13, **c** ~157, and **d** ~3100

extinction is very low, featuring a weakly pronounced peak (in both experimental and simulated spectra) close to 570 nm, which we attribute to multipolar LSP excitation. Due to the changed particles geometry with doubled y-dimension, the dipolar

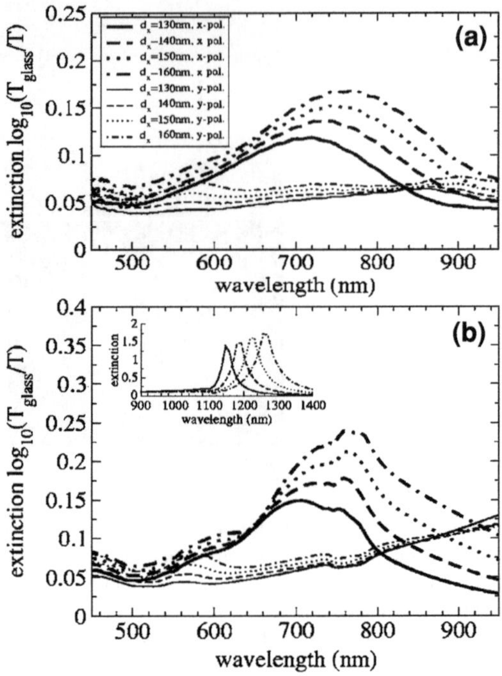

Fig. 5.6 Experimental (**a**) and simulated (**b**) extinction spectra of type C arrays with $\Lambda_x = \Lambda_y = 740$ nm and $d_x = 0.5d_y = 130$, 140, 150 and 160 nm. The *inset* in (**b**) depicts the simulated extinction spectrum for $d_x = 160$ nm and y-polarization in the spectral range from 900 to 1400 nm

LSP peak (that can be excited with *y*-polarization) is shifted to the near-infrared wavelengths (at ~1200 nm) as revealed by simulations in a spectrally extended region (see *inset* in Fig. 5.6b).

5.3.2 TPL Microscopy

The good agreement found between the simulated and experimental extinction spectra encouraged us to apply the developed modeling tool for simulations of the field intensity enhancement and subsequent comparison with the results obtained with the TPL microscopy. For a first comparison, we decided to choose a sample exhibiting the most pronounced effects, i.e., large extinction values and strong polarization influence. Within the range of geometrical parameters studied here, we found the sample having square $160 \times 160\,\text{nm}^2$ nanoparticles but different grating constants $\Lambda_x = 860$ nm and $\Lambda_y = 430$ nm fulfilling best the premises. The experimentally measured extinction spectra as well as the results of FDTD simulations for this array are shown in Fig. 5.7.

The TPL images were recorded at different wavelengths (730, 745, 760, 775 and 800 nm) of the fundamental harmonic (FH) illuminating the sample along with the FH reflection images. Figure 5.8 displays exemplarily the result for the FH wavelength

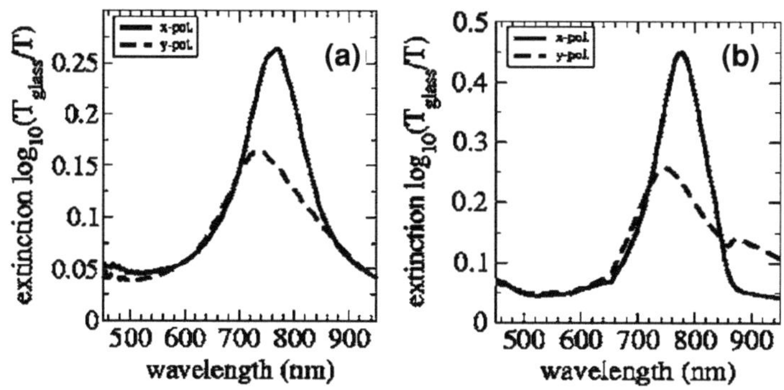

Fig. 5.7 Experimental (**a**) and simulated (**b**) extinction spectra of a gold nanoparticle array with $\Lambda_x = 860\,\text{nm}$, $\Lambda_y = 430\,\text{nm}$, $d_x = d_y = 160\,\text{nm}$

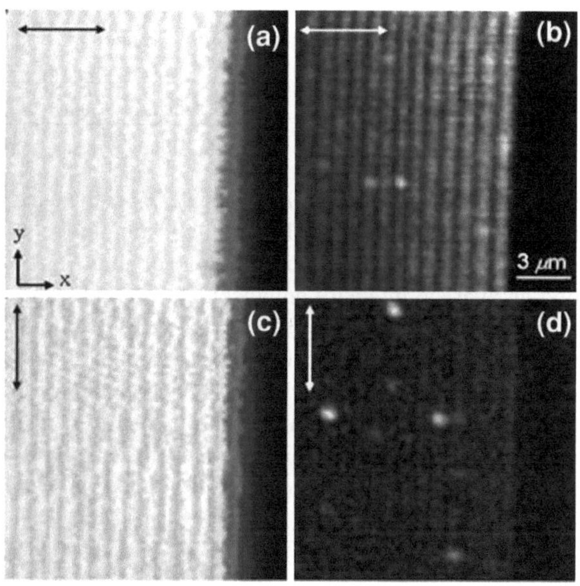

Fig. 5.8 FH (**a,c**) and TPL images (**b,d**) of a gold particle array on glass with $\Lambda_x = 860\,\text{nm}$, $\Lambda_y = 430\,\text{nm}$, and $d_x = d_y = 160\,\text{nm}$ obtained using 3 mW of incident power at the wavelength of 760 nm for the polarizations indicated by *arrows* on the images. The maximum TPL signal is (**b**) ~6200 and (**d**) ~2000 cps

of 760 nm obtained for two polarization configurations of incident FH and detected TPL radiation indicated by arrows on the images (the sample orientation corresponds to that shown in Fig. 5.1b. The image size is $15 \times 15\,\mu\text{m}^2$ with 75×75 points and the incident power was kept at ~3 mW to avoid sample damage.

The FH and TPL images (Fig. 5.8) were obtained starting ~3 μm outside the array of particles. This relatively long distance turned out to be very important in order to get an accurate signal reference from bare glass. The reflection from the glass substrate is rather weak compared to the reflection from gold particles and appears dark in the images. The individual bumps are only resolvable along the x-axis where the separation between them is ~700 nm. This is expected as the resolution in FH and

TPL images has previously been determined to approximately 1 μm and 0.7 μm, respectively [20]. Overall, the FH images appear similar for the investigated polarization directions and wavelengths, while the TPL images are more different. For x-polarized excitation and detection the average TPL signal from the gold particles is relatively high and homogenous, while the y-polarization produces lower average TPL signal but with a few bright spots appearing clearer on the dark background. With respect to the TPL dependence on excitation wavelength, the highest signal was observed for x-polarization and with a resonance around 745 nm. For y-polarization, no pronounced maxima were observed and the signal was considerably weaker (in agreement with the spectroscopy measurements). It should be noted that, in general, the absolute TPL signal levels were found to be very sensitive to the focus adjustment and possible gradual damage of the sample [1].

Based on the method previously used [11, 21], the average (over the particle area) intensity enhancement factor α observed in the TPL measurements can be estimated by comparing the TPL signals from gold nanoparticles to those from smooth gold films. The appropriate relation is given by

$$\alpha = \sqrt{\frac{TPL_{\text{bump}}}{TPL_{\text{film}}} \frac{\langle P_{\text{film}} \rangle^2}{\langle P_{\text{bump}} \rangle^2} \frac{A_{\text{film}}}{A_{\text{bump}}}}, \tag{5.2}$$

where *TPL* is the obtained TPL signal, $\langle P \rangle$ is the used average incident power, and A is the area generating the TPL signal. Using this relation for the TPL-measurements with the A_{bump}-area kept constant, one obtains Fig. 5.9. At the 745 nm resonance we find an average intensity enhancement of $\sim$111 and a maximum (measured at one of the bright spots in Fig. 5.8) of $\sim$250. For sample configuration ($\Lambda_y = 430$ nm) there are two rather than one particle within the focal spot of the exciting laser beam. This larger density of particles can be accounted for in the estimated enhancement by using $A_{\text{bump}} = 2 \times (160\,\text{nm})^2$, resulting in the average intensity enhancement of $\sim$80 and a maximum of 177. This indicates, that the particle near-field interaction of this sample plays only a minor role for the generation of the TPL signal for x-polarization.

5.3.3 FDTD-Results on TPL

In Fig. 5.10 the intensity enhancement values estimated from the simulations are depicted for the wavelength range 450–1100 nm with 20 nm steps. As seen in Fig. 5.10 the simulated TPL intensity enhancement for square particles ($160 \times 160\,\text{nm}^2$) in a rectangular lattice geometry ($\Lambda_x = 860$ nm, $\Lambda_y = 430$ nm) exhibits the same clear polarization dependence as observed in the TPL microscopy measurements (see Fig. 5.9). Furthermore, for both simulated and measured intensity enhancements the ratio between x- and y-polarization is of the same order of magnitude, though the absolute levels are different. The difference in the peak enhancement levels and the slightly shifted to shorter wavelengths peak position (around 750 nm) will be also

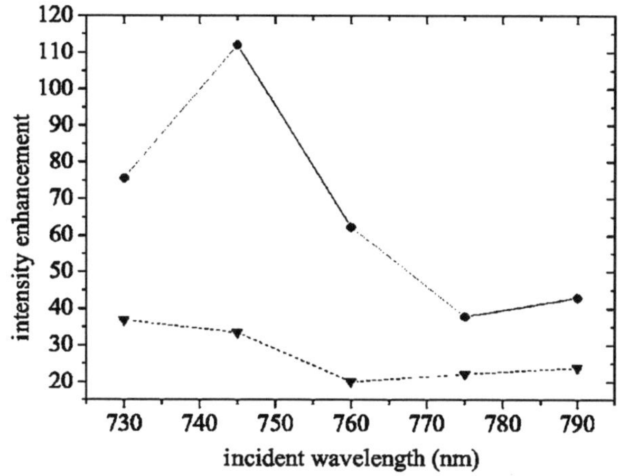

Fig. 5.9 Experimental spectral dependence of the average TPL enhancement (Eq. 5.2) obtained from the particle array on glass. *Solid* and *dashed* lines represent *x*- and *y*-polarization, respectively

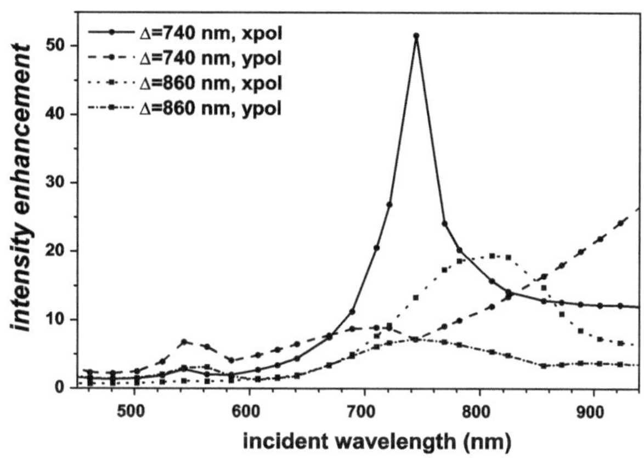

Fig. 5.10 Simulated spectral dependence of the average TPL enhancement obtained from the particle array on glass

explained in Sect. 5.5. The most interesting feature in this context is that, in the simulated field enhancement spectra, one observes a clear red shift of the maximum with the increase of the array period (Fig. 5.10). We found a similar linear shift also for other array periods (not shown). Note that the extinction spectra depend only weakly on the period in this range. Overall, there are indications that the extinction and field intensity spectra of particle arrays might reach maxima at different wavelengths, but this conjecture would require a separate study that can corroborate its existence and explain its origin.

5.4 Spectroscopy and TPL of Au Nanoparticle Arrays on Gold Films

In this section, we apply again the techniques of reflection spectroscopy and TPL microscopy along with the FDTD modeling to study gold nanoparticle arrays placed on a gold surface, aiming at optimizing the array geometry for TPL enhancement and elucidating the roles played by LSP and SPP resonances in local field enhancement effects as well as gaining further insight into the electrodynamical processes involved in TPL generation.

5.4.1 Reflection Spectra

We first consider the far-field reflection spectra of the particle arrays to confirm that the FDTD method is applicable to this system and leads to results in reasonably good agreement with the experiment. Figure 5.11 depicts the experimentally recorded and FDTD simulated reflection spectra for arrays with particles of approximately $150 \times 150 \times 50\,\text{nm}^3$ in a square array with periods $\Lambda_x = \Lambda_y = 740, 780, 820$ and 860 nm. Due to the symmetry of the arrays, the spectra for x- and y-polarization look identical. In very good agreement between experiment and simulations, the reflection spectra display distinct features in three wavelength regions: (A) a single, broad reflection dip at ~550 nm independent of the array period, (B) a dip at the long wavelength wing of the 550 nm dip and (C) a dip (in case of the experimental spectra with a shoulder ~20 nm to the blue of the dip) at a light wavelength close to the value of the array period ($740 - 860\,\text{nm}$). Whereas the independence of dip A on the array period points towards the excitation of a mode localized to the single particles, i.e., the LSP mode, the dip structures B and C clearly depend on the array period and can therefore be related to coupling between the particles. A closer analysis of the spectral dip positions allows to clarify their origin (Fig. 5.12).

Dip C as well as dip B are exactly at the position expected for grating coupling to SPPs on the air-gold interface in the [1 0] and [1 1] mode (the SPP dispersion relation on the array is assumed to follow that of the SPP on an unstructured surface of a 55 nm thick gold film [25]). The experimentally observed shoulder ~20 nm to the blue of dip C in turn is close to the spectral position expected for coupling to grazing grating orders in air. The slight blue shift of its observed spectral position compared to the "ideal" position is due to the far-field emission pattern of dipoles or multipoles above a plane interface, which show considerable strength only close to the grazing angle but are vanishing in the direction parallel to the interface [26]. This shoulder appears much weaker in the simulations and is not discernible in Fig. 5.11b.

To gain information on the parameters which determine the spectral position and strength of the LSP resonance and the grating coupling dips, we produced and investigated samples with different particle shapes and array geometries. For example, rectangular particles created by either bisecting (not shown) or doubling (dash-dotted curves, Fig. 5.13) the y-dimension of the particles lead to a slight splitting of the LSP resonance (peak A) for different polarizations.

For the latter case we observe a red shift (~20 nm) of the resonance for a polarization parallel to the long particle axis (y-polarization, thin dash-dotted curve in Fig. 5.13) and a blue shift (~5 nm) for a polarization parallel to the short particle axis (x-polarization, thick dash-dotted curve in Fig. 5.14). In addition, the experi-

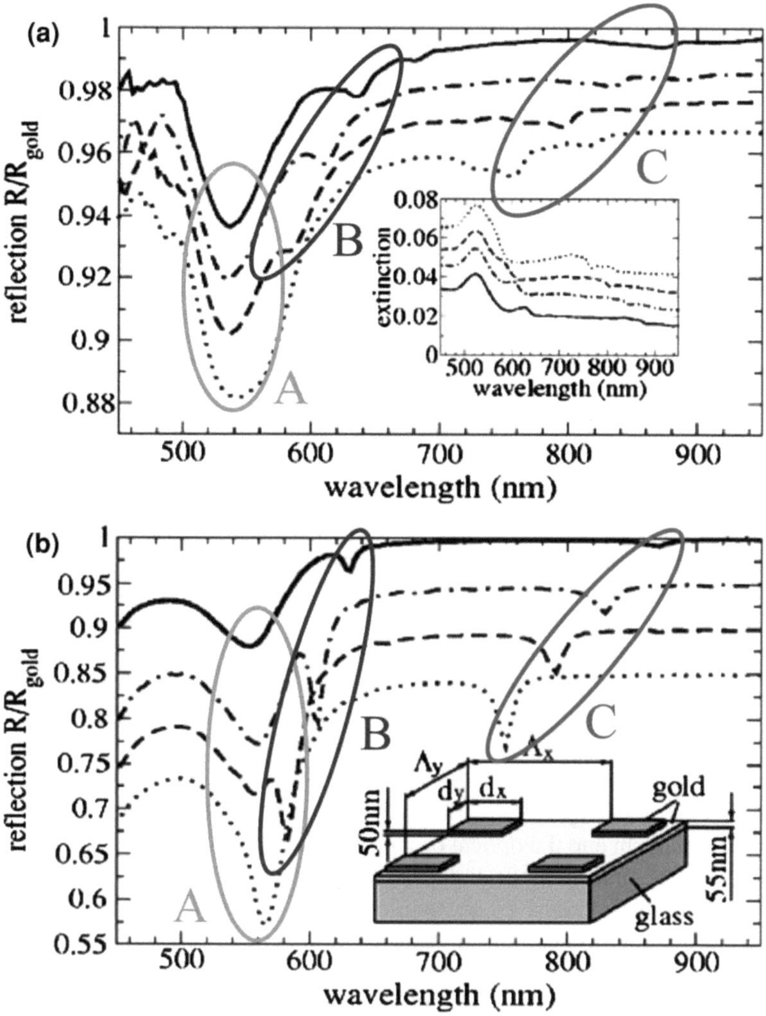

Fig. 5.11 **a** Measured and **b** calculated reflection spectra of particle arrays of square gold particles ($d_x = d_y = 150$ nm, height $d_z = 50$ nm) on a 55 nm thick gold film on *top* of glass (see *inset* part **b**) for different array periods: $\Lambda_x = \Lambda_y = 860$ nm (*solid line*), $\Lambda_x = \Lambda_y = 820$ nm (*dash-dotted line*), $\Lambda_x = \Lambda_y = 780$ nm (*dashed line*), $\Lambda_x = \Lambda_y = 740$ nm (*dotted line*). The curves are vertically offset for clarity. The *inset* in **a** depicts the measured extinction spectra which show features at the same spectral position as observed in the reflection spectra. The spectra are identical for x- and y-polarization

ments for this polarization show a comparably strong occurrence of dip B (grating coupling to the SPP along the [1 1] direction) combined with a slight red shift, for this polarization. For the polarization parallel to the long particle axis (y-polarization) it seems that the efficiency of coupling to grating modes is enhanced while the coupling

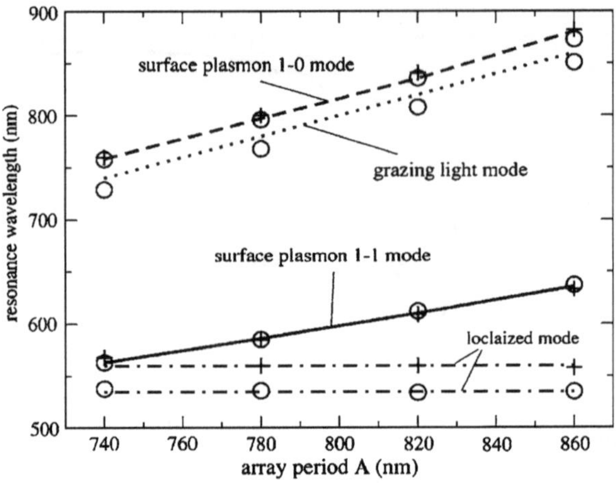

Fig. 5.12 Position of the different dips observed in the reflection spectra versus array period as measured from Fig. 5.11 (*circles* experiment; *crosses* simulation). The *lines* represent the expected dip positions for grating coupling to the [1 1] (*solid line*) and the [1 0] (*dashed line*) surface plasmon modes, the excitation of a grazing light mode (*dotted line*). The horizontal lines reflect the independence of a localized mode on the array period (*dash-dotted lines*)

to the LSP mode gets slightly weaker. These effects are reproduced by the simulations but their origins remain to be clarified and would require further study.

Finally, by bisecting the array period in y-direction ($\Lambda_x = 2\Lambda_y = 860\,\text{nm}$) a complete suppression of the [1 1] SPP modes can be demonstrated (dashed curves, Fig. 5.13). This effect can be qualitatively understood by considering the grating coupling mechanism and the optical properties of the gold film. The SPP wavelength determined by the grating mode and the array period (i.e. 385 nm for the [1 1] mode) requires a frequency that is larger than the onset of the gold *d*-band absorption. SPPs excited within this absorption range are strongly damped and have propagation lengths smaller than their wavelengths and the grating periodicity. Consequently, no resonant grating excitation of these SPP modes can occur.

It is interesting to note the differences in the spectra of these arrays of gold particles on a 55 nm thick gold film to the spectra of similar particles on glass substrate, studied in the last section. The spectra of the particles on glass substrate are dominated by the (shape-dependent) LSP resonance of individual particles, whose strength and spectral width is influenced by the array in terms of suppression of different allowed or forbidden diffraction modes. In contrast, for the particles on gold film, we find a LSP resonance whose spectral position only weakly depends on the particle shape and size (Fig. 5.13) and is always close to ~550 nm. Additionally, we observe different features corresponding to grating coupling to SPPs on the gold-air interface and the excitation of grazingly diffracted light modes in air. The spectral position of these features naturally depends mostly on the array period but their strength can be influenced by the particle shape and size.

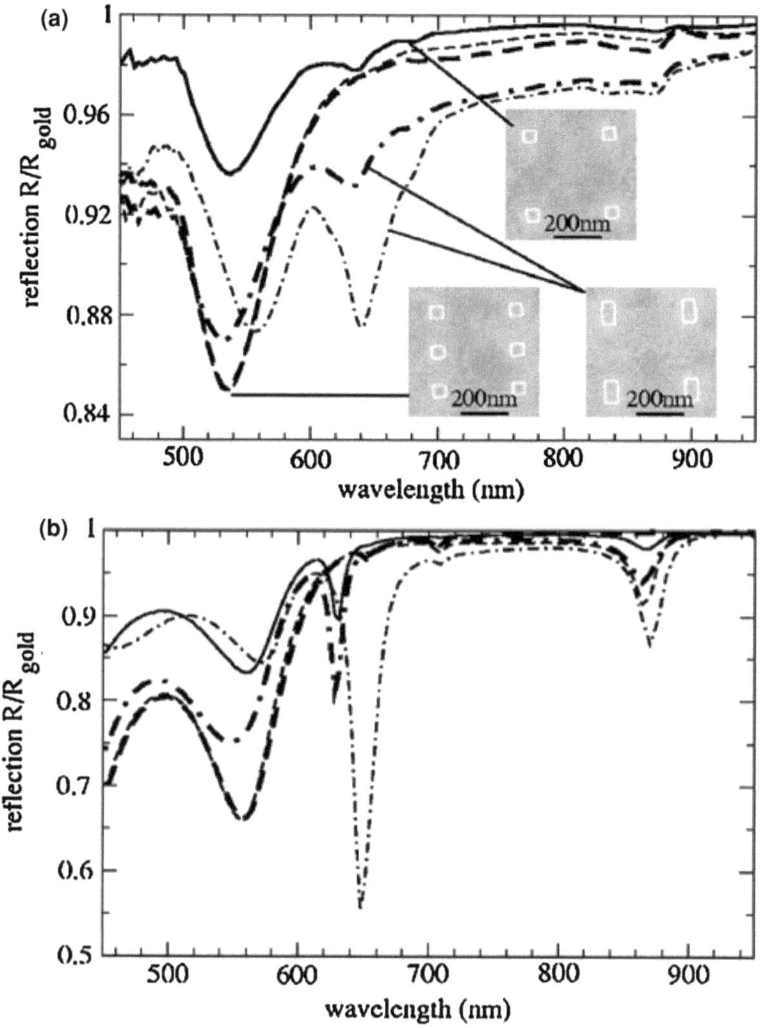

Fig. 5.13 **a** Measured and **b** calculated reflection spectra of particle arrays with $\Lambda_x = \Lambda_y = 860\,\text{nm}$; $d_x = d_y = 150\,\text{nm}$ (*solid line*), $\Lambda_x = 2\Lambda_y = 860\,\text{nm}$ $d_x = d_y = 150\,\text{nm}$ (*thick* and *thin dashed line* for x- and y-polarization, respectively) and $\Lambda_x = \Lambda_y = 860\,\text{nm}$ $d_x = \frac{1}{2}d_y = 150\,\text{nm}$ (*thick* and *thin dash-dotted line* for x-polarization and y-polarization, respectively). The particle height for all arrays is $d_z = 50\,\text{nm}$. The *insets* show electron micrographs of the corresponding arrays

5.4.2 *Optical Near-Field Pattern*

We now turn to the calculated optical near-field pattern and analyze them for the array with $\Lambda_x = \Lambda_y = 740\,\text{nm}$ and particles dimension $d_x = d_y = 150\,\text{nm}$ and height

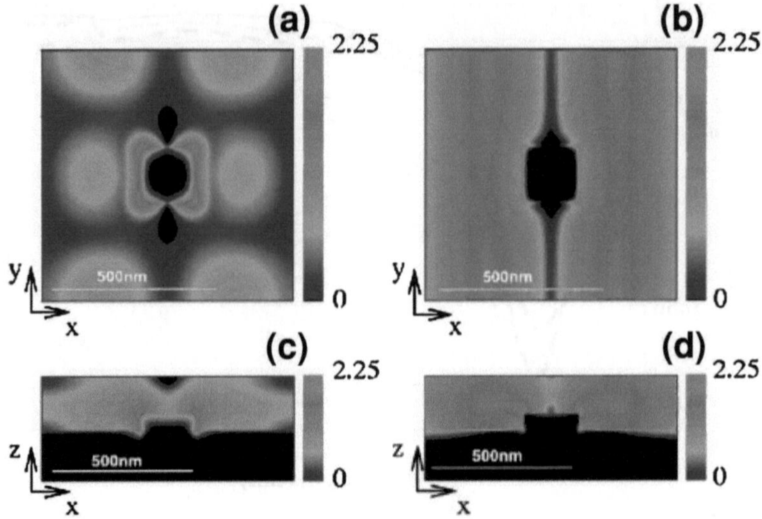

Fig. 5.14 Calculated optical near-field enhancement images for the array with $\Lambda_x = \Lambda_y =$ 740 nm, $d_x = d_y =$ 150 nm (see Fig. 5.11), in the x–y plane at the surface of the gold-film when excited at **a** 563 nm ([1 1] mode) and **b** 752 nm ([1 0] mode). The corresponding images in the *x–z* plane through the particle center are rendered at **c** 563 nm and **d** 752 nm. The plotted quantity is $\log\left(|E(\mathbf{r},\lambda)/E(\lambda)|^2\right)$, where $E(\mathbf{r},\lambda)$ is the electric field amplitude of the array and $E(\lambda)$ is the electric field amplitude in the *top* layer of a flat surface

$d_z = 50$ nm. For this array the LSP resonance and the [11] SPP resonant excitation coincide spectrally, leading to an enhanced SPP excitation and, therefore, a stronger signature in the optical near-fields. We consider first the optical near-field intensities in a *x–y* plane at the surface of the gold film for illumination at the wavelengths of 563 nm (Fig. 5.14a) and 752 nm (Fig. 5.14b), corresponding to the resonant grating excitation of the [1 1] and [1 0] SPP modes, respectively (See Fig. 5.11). The images clearly show standing wave patterns which result from the interference of the excited SPP modes and corroborate the interpretation derived from the spectra. In the first case (excitation at 563 nm), four equivalent SPP modes are excited: [1 1], [1 −1], [−1 1] and [−1 −1]. The interference of these four modes which propagate in the diagonal directions leads to the characteristic pattern observed. Due to the partly longitudinal nature of the SPP field, no SPP modes propagating perpendicularly to the polarization direction of the incoming light can be excited [27]. Therefore, in the case of the excitation at 752 nm (x-polarization), only the [1 0] and [−10] SPP modes are excited whose interference leads to a standing wave pattern with wavefronts parallel to the y-direction, as clearly observed in Fig. 5.14b.

The optical fields are in both cases vertically well-confined to the surface region (Figs 5.14c and d) manifesting thereby their evanescent nature, inherent to SPP modes. In addition to the SPP fields covering a large part of the array surface, strongly localized near-fields are observed close to the upper edges of the particles in both

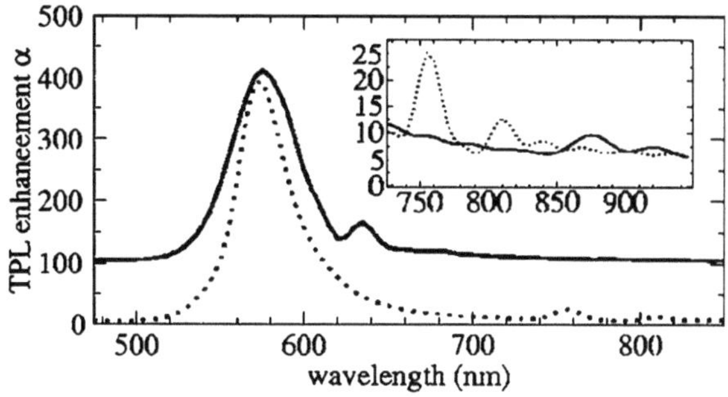

Fig. 5.15 Calculated TPL enhancement spectra for the arrays with $\Lambda_x = \Lambda_y =$ 860 nm $d_x = d_y = 150$ nm (*solid line*), $\Lambda_x = \Lambda_y =$ 740 nm $d_x = d_y = 150$ nm (*dashed line*). For clarity the curves are vertically offset by 100. The *inset* depicts a closeup of the spectral region of 735–950 nm (no offset between curves)

cases (Fig. 5.14c and d). These local field enhancements are due to the lightning rod effect (i.e., field enhancements close to sharp tips or corners) being further enhanced in the first case due to the LSP resonance.

By comparing the near field intensities just below and above the metal surface in the cross-cuts of either Fig. 5.14c or d, one can realize the strong intensity jumps over the gold-air interface in some regions. This is related to the fundamental difference of the continuity condition for the electric field components parallel and perpendicular to the interface. In regions where the electric field is mostly parallel to the metal-air interface, the fields are continuous across the interface, but in regions where the electric field also has a considerable component vertical to the metal-air interface, this component is larger in air by the ratio of the dielectric constants of gold to air (for example at 752 nm excitation $\epsilon_{\mathrm{Au}} \simeq -20.2 + 1.3i$ which can cause a maximum intensity jump of $|\epsilon_{\mathrm{Au}}|^2 \simeq 411$ in case of an electric field purely perpendicular to the interface). This detail highlights the complementary nature of TPL signals, which probe the field inside the metal, versus other methods probing the near field (e.g. surface enhanced Raman scattering or any type of optical near field microscopy) just outside the metal.

5.4.3 TPL Enhancement

The intensity enhancement values estimated from the simulations are shown in Fig. 5.15 in the wavelength range 480–950 nm exemplarily for the arrays with $d_x = d_y = 150$ nm and $\Lambda_x = \Lambda_y = 860$ nm (solid line) and $\Lambda_x = \Lambda_y = 740$ nm (dotted line). The TPL enhancement factor roughly resembles the spectral features in the reflection spectra, i.e. a broad peak (1) at ~575 nm corresponding to the LSP mode but slightly shifted to the red compared to the dip in the extinction spectrum (see Fig. 5.11), and peaks (2) at 635 (solid line) corresponding to the excitation of the [1 1] SPP mode, and (3) at 880 nm (solid line) and 750 nm (dotted line), corresponding to the excitation of the [1 0] SPP mode.

Also for the other arrays investigated up to here (TPL spectra not shown), the major contribution to the TPL signal is predicted to be at ~575 nm. We have noted previously that the LSP resonance wavelengths deduced from the reflection/extinction (far-field) spectra might differ from the TPL enhancement maxima found from near-field calculations (Sect. 5.3). In the current case, the red shift might be due to the circumstance that the reflection dip is associated with the absorption of the resonantly excited LSP mode (and the absorption drastically increases towards shorter wavelengths) whereas the TPL enhancement peaks up at the maximum of the LSP field.

It transpires from the preceding considerations that the insofar investigated arrays are not expected to lead to strong TPL enhancements or significant spectral features in the experimentally accessible spectral range between 730 and 820 nm (Fig. 5.15). Indeed, TPL measurements showed enhancement factors of $\sim$10–20, with spectrally flat characteristics (not shown). For a more valuable comparison of simulated and measured TPL signals, it is necessary to investigate arrays exhibiting pronounced (resonant) TPL features in the spectral range accessible to the experimental setup.

In order to design and fabricate an appropriate sample, we first optimized the array parameters by simulations and found that arrays with $\Lambda_x = \Lambda_y \simeq 750$ nm (similar to the arrays investigated in the previous section) and particle dimensions close to $d_x = d_y \simeq 465$ nm should have a relatively strong resonance associated with the [1 0] SPP excitation in the spectral region relevant for the experiment (Fig. 5.17b). For this sample, the images of the optical near field intensity at the two SPP resonances (580 nm [1 1]-resonance and 800 nm [1 0]-resonance) are depicted in Fig. 5.16, in *x–y* planes 5 nm below and above the gold–air interface, and in the *x–z* planes through the center of the particles. In this case, similar to the smaller particles near field patterns are observed. Again, close to the [1 1]-resonance ($\sim$580 nm), four equivalent SPP modes are excited: [1 1], [1 -1], [-1 1] and [-1 -1], i.e., four SPP waves which propagate in the diagonal directions, leading to a characteristic interference pattern depicted in the *inset* of Fig. 5.16a. In the case of the excitation at a light wavelength of 800 nm, only the [1 0] and [-1 0] SPP modes are excited whose interference leads to a standing wave pattern with wavefronts parallel to the y-direction, which in case of a flat interface would lead to the pattern depicted in the *inset* of Fig. 5.16b. Here, qualitatively similar patterns are observed, but there are also geometrically induced strongly localized near-fields at the particle edges, which considerably contribute to the overall near field intensity and lead to a less obvious appearance of the characteristic [1 1] and [1 0] pattern (Fig. 5.16a, b, c and d). As it will be shown, TPL enhancement is higher for the larger particles as compared to the smaller ones, which is intimately related with the strongly localized near-fields found on them. The reflection spectra of the correspondingly fabricated sample (Fig. 5.17a) exhibit close similarities to the simulations, except for the experimentally observable much stronger occurrence of the dip attributed to the excitation of light scattered at grazing angle to substrate (a close analysis of the simulated spectra reveals also the presence of this feature but as very weak shoulder of the [1 0]-grating coupling dip). This difference as well as the weaker and broader experimentally observed dips compared to the simulations were already observed with the previous samples and can be explained similarly (Sect. 5.5).

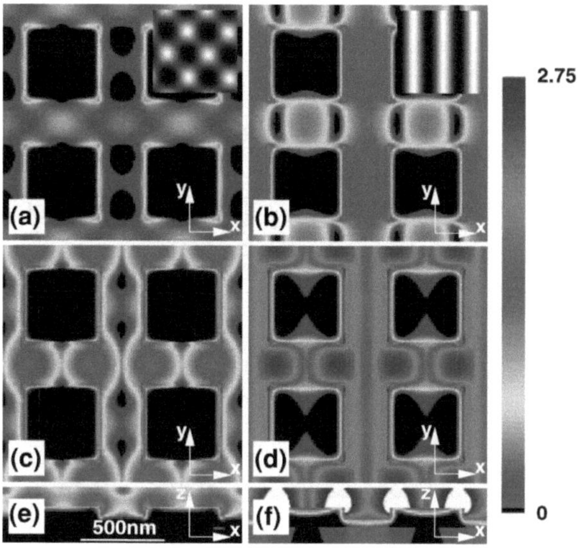

Fig. 5.16 Optical near-field enhancement images (defined as in Fig. 5.14) for the [1 1] (**a**), (**c**) and (**e**) and [1 0] (**b**), (**d**) and (*f*) resonance for the array with $\Lambda_x = 760$ nm, $\Lambda_y = 750$ nm and $d_x = d_y = 465$ nm, in cross-cuts parallel to the substrate, 5 nm below (**a**, **b**) and above (**c**), **d**) the film surface and in cross-cuts perpendicular to the substrate through the center of the particles (**e**, **f**). The *insets* in (**a**) and (**b**) show the interference pattern of similar SPP-waves on a flat interface [18]

TPL spectra from arrays with particle sizes $d_z = 50$ nm and $d_x = d_y = 160$ nm, 265 nm, 364 nm, and 465 nm for polarization parallel to y, and seven different wavelengths (730, 745, 760, 775, 790, 805 and 820 nm) recording reflected FH and TPL microscopy images were measured in the group of Prof. S.I. Bozhevolnyi. The typical FH and TPL images obtained from the area with 465 nm-sized particles are displayed in Fig. 5.18 for the excitation wavelength of 745 nm . For every wavelength, the FH and TPL images were obtained starting ~3 μm outside the array of particles. This relatively long distance was used in order to get an accurate reference from smooth gold surface areas. Note that the FH images have been recorded in the cross-polarized configuration. This means that the smooth gold film (reflecting the FH radiation with the maintained polarization) will appear dark in the FH images, while the gold particles (scattering and changing the light polarization) will appear bright.

Applying the method used in Sect. 5.3, the intensity enhancement factor α observed in the TPL measurements can be estimated by comparing the area averaged TPL signals from the arrays to those from smooth gold films. The used relation is

$$\alpha = \sqrt{\frac{S_{\text{array}}}{S_{\text{film}}} \frac{\langle P_{\text{film}} \rangle^2}{\langle P_{\text{array}} \rangle^2} \frac{A_{\text{film}}}{A_{\text{array}}}}, \tag{5.3}$$

where S is the obtained TPL signal, $\langle P \rangle$ is the used average incident power, and A is the area generating the TPL signal. The average TPL enhancement estimated from the recorded TPL images using this relation is shown in Fig. 5.19 as a function of the FH wavelength for all four investigated samples along with the calculated values of the TPL enhancement. It is clearly seen from the experimental results, that the array with $d_x = d_y = 465$ nm produces the highest average TPL enhancements of ~100, whereas the arrays with smaller particle sizes result in lower enhancements with their

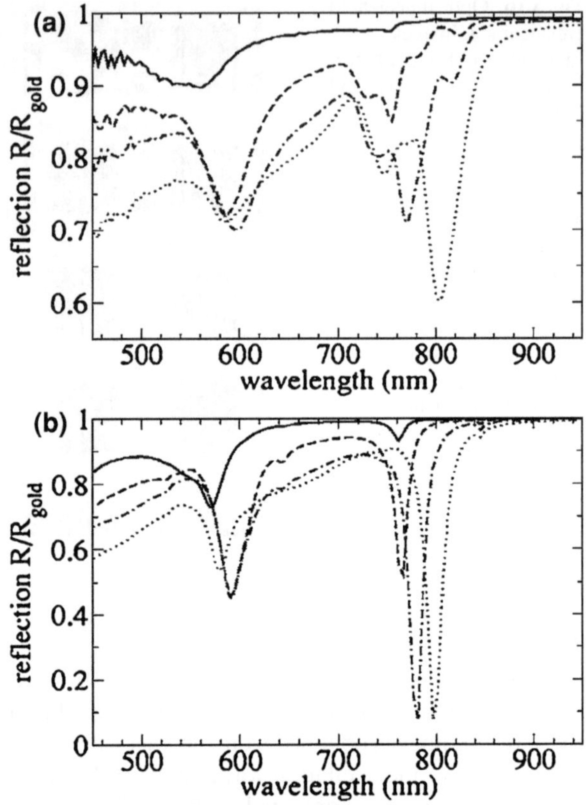

Fig. 5.17 **a** Measured and **b** calculated reflection spectra of particle arrays with $\Lambda_x = 760\,\text{nm}$, $\Lambda_y = 750\,\text{nm}$ and $d_x = d_y = 160\,\text{nm}$ (*solid line*), $d_x = d_y = 265\,\text{nm}$ (*dashed line*), $d_x = d_y = 364\,\text{nm}$ (*dash-dotted line*) and $d_x = d_y = 465\,\text{nm}$ (*dotted line*). The particle height is $d_z = 50\,\text{nm}$ and the polarization is parallel to y

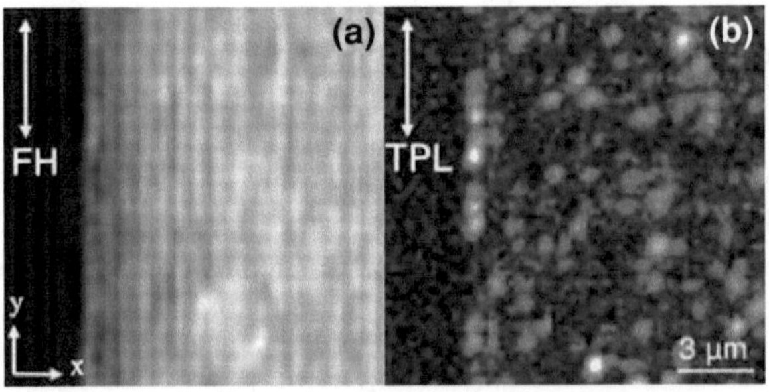

Fig. 5.18 **a** FH and **b** TPL image of a gold particle array with $\Lambda_x = 760\text{nm}$, $\Lambda_y = 750\,\text{nm}$, and particle size $d_x = d_y = 465\,\text{nm}$ and $d_z = 50\,\text{nm}$ obtained using $\sim$0.3 mW of incident power at the wavelength of 745 nm. The maximum TPL signal is $\sim$1600 cps and the polarization of excitation and detected TPL is parallel to y as indicated by *arrows* on the images

peak positions moving towards shorter wavelengths. A qualitatively similar behavior can be observed in the enhancement spectra calculated with the FDTD approach.

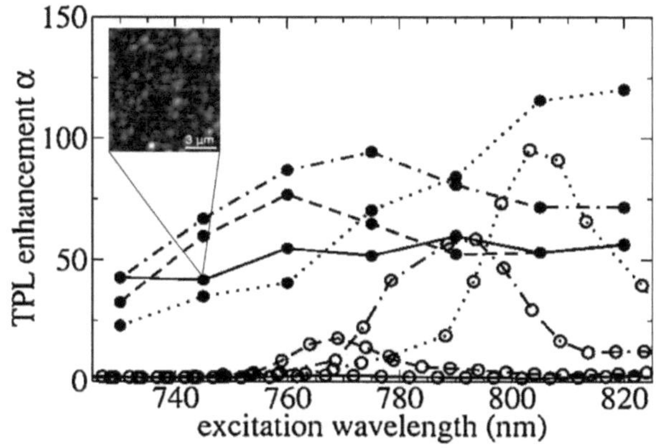

Fig. 5.19 Measured (*filled circles*) and calculated (*open circles*) spectral dependence of the average TPL enhancement (Eq. 5.3) obtained from the particle arrays with $\Lambda_x = 760$ nm, $\Lambda_y = 750$ nm and $d_x = d_y = 160$ nm (*solid lines*), $d_x = d_y = 265$ nm (*dashed lines*), $d_x = d_y = 364$ nm (*dash-dotted lines*) and $d_x = d_y = 465$ nm (*dotted lines*). The particle height is $d_z = 50$ nm. *Inset*: TPL-microscopy image of the latter sample recorded off resonance at 745 nm

It should be mentioned that the maximum TPL enhancement observed from a few individual particles in the array, behaving differently from the average nanoparticles, is ~225. However, at the same time these few particles (bright spots in the TPL images) seem to be more sensitive to damage/reshaping than the remaining particles. Since we aim here at the evaluation of reproducible field enhancements, this damage and reshaping of particularly luminous (individual) positions is neglected in order to allow the excitation power necessary to observe reliable TPL signals from average nanoparticles in the arrays. Note that the incident power used here is between ~0.3 and 0.6 mW for the largest particles ($d_x = d_y = 465$ nm, $d_x = d_y = 364$ nm) and up to ~1.7 mW for the smallest particles ($d_x = d_y = 265$ nm, $d_x = d_y = 160$ nm). These values should be compared to ~3 mW used in the previous TPL measurements from arrays with gold particles on glass.

One can further observe that, except for the smallest particle size, the measured maximum TPL enhancements actually agree with the calculation results within a factor of 2. However, the experimental TPL peaks are broader and less pronounced as compared to the calculated ones, a difference which is consistent with the tendency observed when comparing measured and simulated reflection spectra.

Finally, let us elucidate the issue of spatial confinement of the TPL signals and field enhancement, respectively. In particular, considering the near-field intensity distributions in Figs. 5.14 and 5.16, the question on the effective surface zone responsible for the TPL signal arises, i.e. if it is the particle alone which emits the TPL signal. This issue can be clarified by plotting $\rho = S_{\text{part}}/S_{\text{tot}}$, the relative contribution of the simulated TPL signal originating from the particle surface only (S_{part}) to the simulated total TPL signal (S_{tot}) as a function of the excitation wavelength (Fig. 5.20).

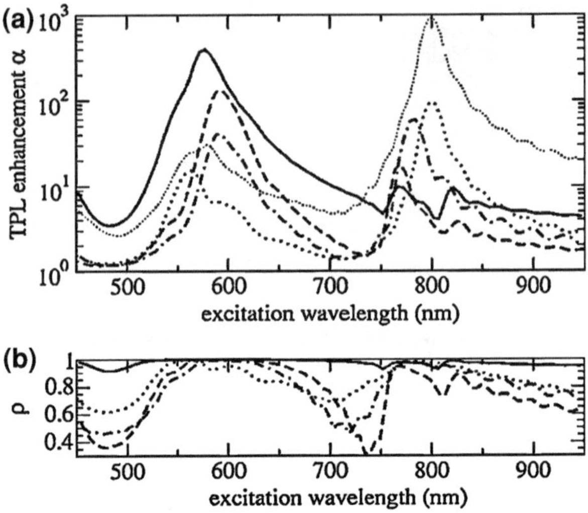

Fig. 5.20 **a** Average TPL enhancement α and **b** relative contribution ρ of the particle area to the overall TPL signal for arrays with $\Lambda_x = 760$ nm, $\Lambda_y = 750$ nm and $d_x = d_y = 160$ nm (*solid line*), $d_x = d_y = 265$ nm (*dashed line*), $d_x = d_y = 364$ nm (*dash-dotted line*) and $d_x = d_y = 465$ nm (*dotted line*). The particle height for all arrays is $d_z = 50$ nm and the polarization is parallel to y. The *thin dotted line* in part **a** depicts the enhancement factor α calculated on the air side of the gold-air interface of the array with $d_x = d_y = 465$ nm

It transpires from the computed TPL signal that, although the peak at about 800 nm comes from the SPP excitation (i.e., related to delocalized SPP fields), the TPL originates primarily from the particles whenever the enhancement factor α is of considerable strength. The reason for this is that the excited SPPs provide additional "illumination" of the particles, contributing thereby to the formation of strong fields inside the particles (particularly around the edges) which are then responsible for the TPL and near field enhancement inside the gold. To emphasize the relation of the near field enhancements on both sides of the gold-air interface, we additionally depict for comparison in Fig. 5.20 the unit-cell average of the enhancement factor α calculated over a layer just *above* the gold surface (which reflects e.g. the gain in surface enhanced Raman scattering) for the array with $d_x = d_y = 465$ nm (thin dotted curve). As can be seen in the graph, it roughly reproduces the general shape of the average TPL enhancement below the gold surface (dotted curve), but is larger by one order of magnitude. As discussed in detail in Sect. 5.4.2, this is due to the continuity relations, which require a jump of the electric field component perpendicular to the surface.

5.5 Confrontation of Simulations to Experiments

Despite the good qualitative agreement between simulations and experiments, there are also significant deviations. The observed differences between the experimental and simulated reflection spectra: Figs. (5.2, 5.3, 5.6, 5.7) (gold on glass) and (5.11, 5.13, 5.17) (gold on gold), and the TPL-enhancement spectra on glass Figs. (5.9, 5.10) and on gold Fig. 5.19, may originate from several reasons: (1) experimental variations in particle size and array period, (2) the finite numerical aperture of the spectrometer setup (not considered in the simulations), (3) the TPL excitation and detection geometry, (4) deviations of the dielectric function of gold between the actual sample and the Drude-Lorentz fit used for FDTD simulations, including non-locality effects and spatial variations, (5) surface roughness (not considered in the simulations), and (6) influences of the FDTD boundary conditions and the finite array size. A detailed consideration of the different possibilities leads to the following estimations of the different contributions.

1. Experimental variations in the particle size and array period: Due to the fabrication tolerances, the geometrical parameters of the samples investigated may vary in the order of ~10 nm, i.e., by 2%. This causes small phase mismatches in case of the grating coupling to SPP and related to that a weakening and broadening of the corresponding resonances. The changes in the peak position caused by such variations are at maximum ~15 nm and can therefore partly account for the experimentally observed broader peaks. However, for the localized resonance at ~520 nm on gold particles on gold, these variations are not sufficient to explain the larger experimental observed peak width in the reflection-loss spectra, since this peak is basically independent on variations of the particle shape in this range.
2. The finite numerical aperture of the spectrometer setup leads to an angular spread of the incident light. At inclined incidence of a plane wave, the reflection-loss peaks caused by grating coupling to SPP-modes split into a blue- and a red-shifted contribution. For an angular spread corresponding to NA = 0.075 as in the experiments this would cause a peak broadening of at maximum ~120 nm at a light wavelength of ~800 nm, depending on the effective NA of the illumination path (angular intensity distribution).
3. From the experimental point of view, the TPL excitation and detection geometry sources of disagreement between experiments and theory might be justified as follows. The observed difference with the simulations might partly be due to the fact that the TPL measurements use a tightly focused beam with a (correspondingly) wide angular spectrum and a small spot-size of only ~1 μm. This can result in both, a broadening of the peaks and an increase of the background, by facilitating for example, SPP excitation at about any wavelength in the wavelength range (contrary to what one has in simulations). Moreover, the TPL radiation originating from gold areas with strong field enhancements has unknown angular distribution and interacts with the scattering system (i.e., particle array), so that the detected TPL is in fact also subject to all scattering phenomena (scattering at surface roughness, coupling to SPP resonances,...) considered above for the illuminating radiation. However, for coupling to SPP modes we do not expect any relevant influence of the grating, since the propagation length of SPP's at the spectral range below 550 nm is too low to lead to any grating effects with the array periods consider here [28]. Additionally, the experimental results are

affected by the circumstance that the rather weak TPL signals exhibit considerable uncertainties, especially for longer wavelengths, due to inaccuracy in the focus adjustment, possible gradual damage of the sample, etc. Finally, the fact that the TPL enhancement levels measured far from the resonances (for instance the $d_x = d_y = 160$ nm-particle array in Fig. 5.19) do not approach unity should be related to the TPL response from corrugated surfaces (here, due to the surface processing when fabricating particles), which will always be larger than from the flat surface.From the theoretical standpoint, we expect TPL emission being proportional to $|E|^4$. But it is not clear how to count how many photons will be able to leave the metal and reach the detector, and how the probability for this process depends on both the direction of the electric field and the depth at which this process takes place. So a proper comparison with experiments may be much more complex. For instance, results could depend on whether the TPL emission comes from inside the metal, so it should be somehow considered in theory. Clearly, this should bring into account geometrical factors related to the emission and detection of TPL photons; for instance, are photons emitted close to the bump base detected? Therefore, the field inside the whole metal should be included, specially as TPL is emitted at shorter wavelengths, for which the skin depth is larger than for the excitation wavelength. Without a better theory for the TPL enhancement, it would be worth to estimate the importance of this factors by calculating separately the contribution from different places of the structure. Figure 5.21 shows Yee's cell (Sect. 1.2.1). In this cell, different electric field components are represented in different spacial points. We have chosen the parallel components to be represented at the interface, while the cell perpendicular components are chosen to lie either inside or outside. In Fig. 5.22, we revise a result already shown in Fig. 5.10. In this figure, the curve depicted with square symbols is that the E-fields are estimated just outside the top surface of the metal particle. Circular symbols show what we obtain only considering the top surface just inside the metal (E_x and E_y at the interface and E_z evaluated at 2.5 nm depth inside the metal, as the mesh size is 5 nm). The latter estimation for α is smaller than the former, as the E_z component is a factor $\varepsilon(\omega)$ smaller than in the previous calculation. The rest of the curves are estimations for α considering the fields inside the metal up to the the distance to the top surface indicated by the labels. Notice that in these calculations we include the contribution from the side metal interfaces. The integration is up to a depth of 37.5 nm, since we do not expect the TPL signal reaching the metal surface if it is generated at the bottom. In any case, there is still a strong polarization dependence in the calculated α, for the different depths chosen. Figure 5.23 renders the same plot as Fig. 5.22, but for the other structure considered in Fig. 5.10. In this case, α is much smaller when E_z contributes from inside than when it does from just outside. The point is that if the electric field is mainly along the z direction, on the top surface it is decreased by a factor $\varepsilon(\omega)$, but on the lateral sides it is not. This is the case here, and the main contribution comes from a "belt" of high E_z at about half depth of the metal bump. Here the variation with integration depth is very fast. In the case of gold particles on gold, the key point is that α, when TPL contributions are taking into

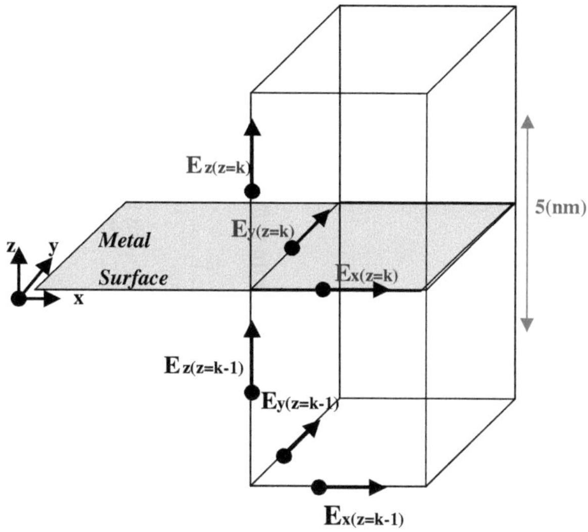

Fig. 5.21 Yee's cell

account inside the metal, gives a result close to what we obtain without doing that, but only for the particle resonances. The "pure SPP" peaks appearing are no longer so strong, as the fields in this case have mainly E_z component.

Deviations of the dielectric function of gold between the actual sample and the Drude-Lorentz fit used for FDTD-simulations: For the FDTD-simulations, the frequency dependent data of the experimentally measured dielectric function of gold are fitted with a Drude-Lorentz behavior (see Sect. 1.2.4). This leads to a very good approximation at larger wavelength, but increasing mismatch with decreasing wavelength in the range below 500 nm. This mismatch can account for deviations between simulated and measured reflection spectra at shorter wavelength and could be reduced by e.g. adding a second Drude-term to the fit. Additionally, the effective dielectric function of the gold-film and gold-particles could deviate from the literature values (e.g., due to surface morphology, see below).

5. Surface roughness: By a detailed analysis of the SEM image of the gold particle arrays on gold (Fig. 5.13a) on can realize a structural difference between the polycrystalline gold film and particle surfaces, i.e., there are smaller grains (∼20 nm in diameter and ripples on the particles compared to larger crystallites in the size range of ∼50–500 nm on the film outside the particles. This qualitative difference in the gold nanostructure comes from the surface processing when fabricating the particles and might cause changes in the effective dielectric function, especially an increase of the imaginary part due to enhanced surface scattering. This contributes to the less pronounced, weaker peaks in the experimental reflection and TPL spectra. Additionally, the surface corrugations lead to additional localized resonances, which are best visible in TPL images recorded off resonance (inset Fig. 5.19) by the randomly distributed bright spots, whose positions depend

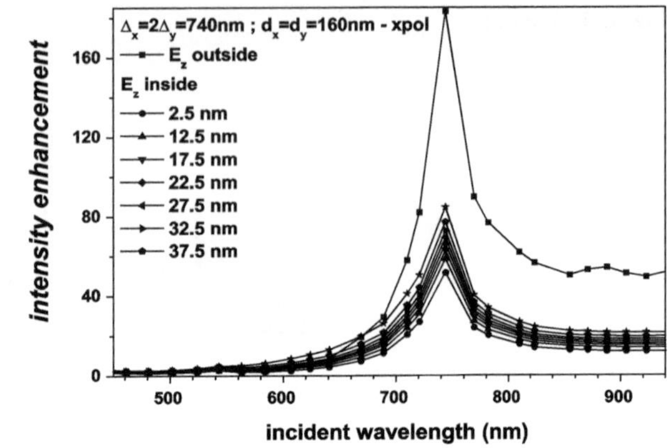

Fig. 5.22 Estimation of the integrated TPL enhancement up to different "depths" inside the metal surface, obtained from particle arrays on glass ($\Delta_x = 2\Delta_y =$ 740 nm, $d_x = d_y = 160$ nm, x-pol)

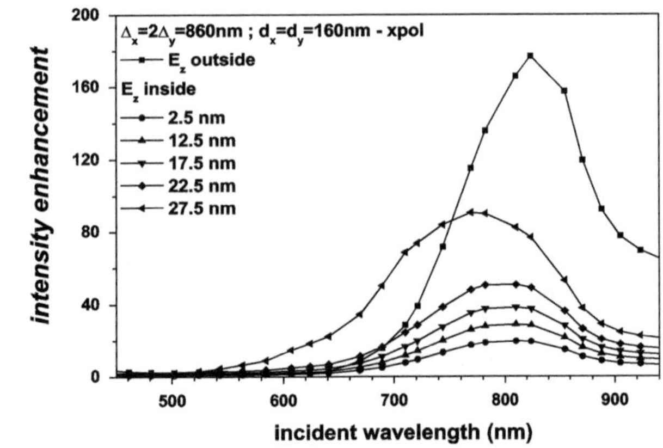

Fig. 5.23 Estimation of the integrated TPL enhancement up to different "depths" inside the metal surface, obtained from particle arrays on glass ($\Delta_x = 2\Delta_y =$ 860 nm, $d_x = d_y = 160$ nm x-pol))

on the excitation wavelength. In average over a larger area, these resonances do not lead to spectrally confined features, but are responsible for the offset in measured TPL compared to simulations and for the fact that the TPL enhancement levels measured far from the resonances (especially for the $d_x = d_y = 160$ nm-particle array) do not approach unity.

6. Influences of the boundary conditions: On the vertical walls of the unit cells, the simulations consider strictly periodic boundary conditions, i.e., infinite arrays, but the experimentally investigated arrays are certainly finite ($100 \times 100\,\mu\text{m}^2$). However, since the propagation length, and therefore the interaction distance between the particles ($\sim$20 μm at 800 nm wavelength), is much smaller than the arrays size, we do not expect relevant modifications of the results. On the bottom and top boundary of the volume considered in the FDTD simulations, absorbing boundary conditions realized by a combination of a UPML and a CCOM layer (Sect. 1.2.5) at a distance of $\sim$0.6 μm above and below the gold film are applied. This is to better absorb the energy flowing at grazing incidence, which is of special importance in the considered system, as small SPP peaks in reflection have to be

resolved and small errors in the reflection of grazing modes due to unwanted lack of absorption by the absorbing layers could be attributed spuriously to SPP resonances. Since this combination of UPML and CCOM is carefully chosen and tested for the simulations, we also do not expect artefacts in the spectra or near fields arising from these boundaries.

5.6 Conclusions

In conclusion, we have investigated the electrodynamical processes involved in the generation of TPL from arrays of rectangular gold nanoparticles deposited either on a glass substrate or on a thin gold film. FDTD simulations have been combined with linear extinction spectroscopy and TPL-microscopy to gain insight into this particular problem.

For gold nanoparticles on glass, the simulations show pronounced effects when the particle resonances spectrally overlap with array resonances. Such effects are not well captured by the experiments. We attribute this to the geometrical imperfections of the samples and the measurement process. TPL enhancements were found to be in the range of 10^2 with a sharp spectral response. The FDTD calculations reproduce well the experimental TPL excitation spectra considering E^4 integrated over the "top" layer ($\sim$half a skin depth) of the particles as the origin of the TPL signal. Additionally, we found indications that the spectral position of the maximum near field intensity enhancement might differ considerably from the position of the maximum seen in the extinction spectra, depending on the period of the particle array.

In the case of gold nanoparticles on top of a thin gold film, the dimensions of the nanoparticles and the array periods were systematically varied to optimize the strength of the SPP resonance in the wavelength range accessible to the experimental characterization techniques. On the optimized array a TPL enhancement up to $\sim$200 has been observed with a relatively broad spectral response. It could be demonstrated that TPL enhancement is well described by our simulations also for this configuration, where we assume that it is related to the field intensity enhancement just below the gold surface, i.e., inside the gold. We could show, that even if the optimized resonance at $\sim$800 nm is due to a resonant excitation of a delocalized SPP mode, the maximum field enhancement (an thereby the origin of the TPL signal) is localized at the particles. This is due to a combination of geometrical field enhancement (lightning rod effect) and the better penetration of the field into the metal at the particle edges. Additionally, our simulations reveal, that the enhancement factor calculated just outside the gold (as it would be probed by e.g. surface enhanced Raman scattering) is in average one order of magnitude larger than inside the gold. The origin of this can be found in the continuity relation across the gold-air interface, which requires the electric field component perpendicular to the interface to be enhanced by the ratio of the dielectric functions.

In addition, by a careful comparison of experimental results versus FDTD simulations, we have identified the parameters responsible for the differences between experiment and theory. To overcome these differences between simulations and experiment, an improved experimental control of surface structure and crystallinity of the gold film and particles and a better knowledge of the gold dielectric function are crucial. For the simulations the possibility include also surface roughness would lead to a substantial improvement. However, the generally reasonable well agreement of simulations and experiments can be interpreted in a way, that macroscopic Maxwell equations as solved by the FDTD code are suitable for a detailed description of similar systems.

References

1. P. Muhlschlegel, H.-J. Eisler, O.J.F. Martin, B. Hecht, D.W. Pohl, Science **308**, 1607 (2005)
2. S. Emory S.R. Nie, Science **275**, 1102 (1997)
3. J. Azoulay, A. Débarre, A. Richard, P. Tchénio, J. Microsc. **194**, 486 (1999)
4. M. Righini, P. Ghenuche, S. Cherukulappurath, V. Myroshnychenko, F.J. García de Abajo, R. Quidant, Nano Lett. **9**, 3387–3391 (2009)
5. U. Kreibig, M. Vollmer, *Optical Properties of Metal Clusters (Springer Series in Materials Science No 25)* (Springer, Berlin, 1995)
6. M. Moskovits, Rev. Mod. Phys. **57**, 783 (1985)
7. J.R. Lakowicz, Anal. Biochem. **298**, 1 (2001)
8. J.R. Lakowicz, Y. Shena, S. D'Auria, J. Malickaa, J. Fangb, Z. Gryczynskia, I. Gryczynskia, Anal. Biochem. **301**, 261 (2002)
9. N. Félidj, J. Aubard, G. Lévi, J.R. Krenn, G. Schider, A. Leitner, F.R. Aussenegg, Phys. Rev. B **66**, 245407 (2002)
10. B. Lamprecht, G. Schider, R.T. Lechner, H. Ditlbacher, J.R. Krenn, A. Leitner, F.R. Aussenegg, Phys. Rev. Lett. **84**, 4721 (2000)
11. P.J. Schuck, D.P. Fromm, A. Sundaramurthy, G.S. Kino, W.E. Moerner, Phys. Rev. Lett. **94**, 017402 (2005)
12. A. Mooradian, Phys. Rev. Lett. **22**, 185 (1969)
13. G.T. Boyd, Z.H. Yu, Y.R. Shen, Phys. Rev. B **33**, 7923 (1986)
14. V.M. Shalaev, C. Douketis, T. Haslett, T. Stuckless, M. Moskovits, Phys. Rev. B **53**, 11193 (1996)
15. M.R. Beversluis, A. Bouhelier, L. Novotny, Phys. Rev. B **68**, 115433 (2003)
16. A. Hohenau, J.R. Krenn, J. Beermann, S.I. Bozhevolnyi, S.G. Rodrigo, L. Martin-Moreno, F. Garcia-Vidal, Phys. Rev. B **73**, 155404 (2006)
17. A. Hohenau, J.R. Krenn, F.J. Garcia-Vidal, S.G. Rodrigo, L. Martin- Moreno, J. Beermann, S.I. Bozhevolnyi, Phys. Rev. B **75**, 085104 (2007)
18. A. Hohenau, J.R. Krenn, F.J. Garcia-Vidal, S.G. Rodrigo, L. Martin- Moreno, J. Beermann, S.I. Bozhevolnyi, J. Opt. A: Pure Appl. Opt. **9**, S366 (2007)
19. M.A. McCord, M.J. Rooks, *Handbook of Microlithography Micromachining and Microfabrication vol 1* (WA: SPIE and the Institute of Electrical Engineers, Bellingham, 1997)
20. J. Beermann, S. Bozhevolnyi, K. Pedersen, J. Fage-Pedersen, Optics Commun. **221**, 295 (2003)
21. J. Beermann, S.I. Bozhevolnyi, Phys. Status Solidi C **2**, 3983 (2005)
22. G.T. Boyd, T. Rasing, J.R.R. Leite, Y.R. Shen, Phys. Rev. B **30**, 519 (1984)
23. D. Fromm, A. Sundaramurthy, P. Schuck, G. Kino, W. Moerner, Nano Lett. **4**, 957 (2004)

24. W. Rechberger, A. Hohenau, A. Leitner, J.R. Krenn, B. Lamprecht, F.R. Aussenegg, Optics Commun. **220**, 137 (2003)
25. J.J. Burke, G.I. Stegeman, T. Tamir, Phys. Rev. B **33**, 5186 (1986)
26. H.F. Arnoldus, J. Opt. Soc. Am. A **22**, 190 (2005)
27. J. Krenn, H. Ditlbacher, G. Schider, A. Hohenau, A. Leitner, F. Aussenegg, J. Microsc. **209**, 167 (2003)
28. J.A. Dionne, L.A. Sweatlock, H.A. Atwater, A. Polman, Phys. Rev. B **72**, 075405 (2005)

MIX
Papier aus verantwortungsvollen Quellen
Paper from responsible sources
FSC® C105338

If you have any concerns about our products,
you can contact us on
ProductSafety@springernature.com

In case Publisher is established outside the EU,
the EU authorized representative is:
Springer Nature Customer Service Center GmbH
Europaplatz 3, 69115 Heidelberg, Germany

Printed by Libri Plureos GmbH
in Hamburg, Germany